ISBN 978-1-332-12794-8
PIBN 10288455

This book is a reproduction of an important historical work. Forgotten Books uses
state-of-the-art technology to digitally reconstruct the work, preserving the original format
whilst repairing imperfections present in the aged copy. In rare cases, an imperfection in
the original, such as a blemish or missing page, may be replicated in our edition. We do,
however, repair the vast majority of imperfections successfully; any imperfections that
remain are intentionally left to preserve the state of such historical works.

1 MONTH OF
FREE
READING

at

www.ForgottenBooks.com

By purchasing this book you are eligible for one month membership to ForgottenBooks.com, giving you unlimited access to our entire collection of over 700,000 titles via our web site and mobile apps.

To claim your free month visit:

www.forgottenbooks.com/free288455

Similar Books Are Available from
www.forgottenbooks.com

FIRES AND
IRE-FIGHTERS

*A History of Modern Fire-Fighting
with a Review of Its Development
from Earliest Times*

BY
JOHN KENLON
CHIEF OF NEW YORK FIRE DEPARTMENT

WITH ILLUSTRATIONS FROM
PHOTOGRAPHS

NEW YORK
GEORGE H. DORAN COMPANY

TH 9117
K4

DEDICATED
TO
MY COMRADES

THE MEMBERS OF
THE INTERNATIONAL ASSOCIATION
OF FIRE ENGINEERS

270947

ACKNOWLEDGMENTS

The Author desires to make special and grateful acknowledgment,

To Mr. Alan Lethbridge for his gracious and valuable assistance in the preparation of this book.

To Mr. Wilbur E. Mallalieu, General Agent of The National Board of Fire Underwriters for help in the preparation of statistical tables and for the use of some special photographs.

To Mr. Edwin O. Sachs and his co-directors in the British Fire Prevention Association for permission to use material and pictures from the reports of their Association.

To Mr. Frederick Smythe for the use of photographs.

CONTENTS

vii

CONTENTS

ILLUSTRATIONS

ILLUSTRATIONS

FIRES AND FIRE-FIGHTERS

CHAPTER I

A COMMON axiom amongst fire-fighters is that no definite rules can be formulated, which wholly embody the principles of their craft. It is argued that since no two fires are absolutely alike in all respects, that which would be efficacious in one instance would be absolutely futile in another. This proposition is fallacious. Physicians might just as well advance the theory that since no two individuals are constitutionally alike, it is useless to apply the same treatment for some well known disease, even with those modifications necessitated by physical differences. Of course this is a "reductio ad absurdum," since doctors study their patients scientifically, following general principles resulting from experience, only varied in minor details according to the exigencies of the case. Similarly, notwithstanding differences in construction and occupancy, it is perfectly feasible to fight fires with intelligence born of systematic acquaintance with certain fixed data, and it may be added, with some degree of scientific exactitude. As there, are prime factors in the treatment of illness, particularly if it be contagious, such as the removal of the patient to a place where it is almost impossible for the disease to be communicated to others, so it is with fire. The first general principles to be observed include naturally the confinement of an outbreak to as narrow a space as possible, the safety of contiguous property, the prevention of loss of life and the centralization of the outbreak as a whole. To this must be added the concentration upon the point of greatest danger of all the forces at the command of the officer in charge. In the following chapters an attempt has been made to deal with this subject in such a manner that while the professional fire-fighter shall find much information which will be of value to him, the lay reader shall likewise discover material for thought, as well as food for the imagination.

It has been estimated that no less than 64 per cent. of all fires occur in the homes of the people, and though these may not be attended by the tremendous financial losses consequent upon outbreak in warehouses, office buildings and the like, they strike fear into the heart in a greater degree, for it is the human hazard which is at stake.

Few realize, also, the unremitting labor, the devotion to service, the daily acts of heroism, the mental and physical strain, and the inadequate acknowledgment in many instances by the public of the achievements of the genus fireman. Not that he wishes to be advertised, but since the soldier, the sailor and even the policeman loom large in general estimation, it seems only just that something should be written illustrative of the responsibilities entrusted to his charge. To how many people does it ever occur that negligence on the part of a policeman may result in the loss by robbery of a few thousand dollars or the sacrifice of at most two or three human lives by murder; while the same fault on the part of a fireman may entail some hideous disaster involving scores of lives or the loss of millions of dollars. Further, is it realized, that whereas the soldier or sailor risks his life for his country at rare intervals, the fireman takes the same chances regularly in the course of his daily avocation.

Thus it will be seen that no occupation or career should make greater appeal to the sympathy and interest of the public than that of the firemen who constitute a force which stands for much and without which the insecurity of life would be increased tenfold.

In addition, the advance of science and the evolution of the simple building into the highly complex structure necessitated by modern requirements have in their turn caused a corresponding advance in the theory and practice of fire-fighting. Questions of such import as the alleviation of congestion in crowded districts, the provision of suitable accommodation for domestic and business premises and the supply of the minimum of light and air compatible with modern ideas of hygiene, have led architects to find their only solution in the piling up of story upon story till, with the Woolworth building in New York, realms of space hitherto unpierced except by the Eiffel Tower have surrendered to their all-conquering demand. And finality has by no means been reached in this direction. No wonder, therefore, that those responsible for fire control have paused, perplexed momentarily at the problems confronting them. Generally speaking, except under the rarest of circumstances, it is only possible to fight fires from the street up to a height of seven stories, after that reliance must be placed upon the fire appliances within the building coupled with the tactical skill of the firemen in using the same. This is one of the instances in which the scientific training of a fire department is manifested. The isolation of elevator shafts, the prevention of flames being drawn from floor to floor through windows and the avoidance of that most dangerous enemy, back draught, constitute features of enormous significance.

Similarly fire apparatus has grown in complexity and its handling requires a corresponding degree of judgment and skill. The old days of the manual have gone forever, and though for many centuries little advance was made in the mechanical aspect of fire-fighting equipment, the last fifty years have witnessed a complete revolution in the means and methods employed. As the hand drawn "manual" gave way to the horse drawn steam engine, so has the latter in its turn been succeeded by the automobile gasoline pump. Likewise the Roman ladder, which for years marked the limit of human ingenuity as applied to means of entry to and rescue from burning buildings, has been superseded to a large extent by mechanically operated extension ladders of great length. Such apparatus as water towers, searchlights, high pressure pumps, dangerous structure traps and so forth, presuppose a high degree of scientific skill and technical knowledge on the part of the fire-fighter, who may thus legitimately claim to belong to a well defined profession. Since appliances vary in different parts of the world, according to local needs, the author has included in this volume some slight account of the equipment and methods of foreign departments, which would prove serviceable for purposes of comparison. Equally, full descriptions will be found of the most modern mechanical devices in use, which it is hoped will be of real service to those who are interested in the subject from a practical standpoint. There is no doubt that at last the world has awakened to the economic importance of fire control. Insurance risks have become so stupendous that those involved financially in the same, demand the acme of scientific foresight and the maximum of human enterprise towards the protection of their capital. It is true that in some quarters there is a regrettable tendency to gamble on fire risks, which brings in its train sporadic outbursts of incendiarism, whereby, in many cases, human lives are jeopardized. But with the exposure of such dubious modes of increasing business and with a realization of their results, it seems beyond question that saner and wiser counsels will prevail.

These are days of keen competition as applied to the search after a bare livelihood and the pay and prospects of the fireman are such that they well merit the attention of young men with ambition and brains. The life is a healthy, if strenuous one, while the position of Fire Chief, at any rate in America, is within reach of all comers and the goal one to be envied. Should this work prove to be the means of encouraging the right type of man to come forward, then the writer will be happy in the knowledge that his labour has not been in vain.

It will be noticed that a chapter has been devoted to a consideration of how best to deal with fires in private houses and the most prolific causes of these outbreaks. Carelessness can never be wholly eradicated from human nature, but this same failing is one of the prime factors constituting the fire risk of the citizen. Not long since a guest in a hotel thoughtlessly threw away a lighted cigarette end into a waste paper basket. In due course the contents burst into flames, set alight the curtains, and eventually involved the whole floor of the building, causing incidentally the loss of three lives. That same story is repeated week by week and day by day the world over, and yet the lesson never seems to be appreciated. Hence, the next best thing to prevention being cure, an attempt has been made in the chapter indicated to formulate certain simple rules which if followed will go a long way towards controlling the blaze until such time as professional help shall arrive. Further, it is not generally realized by what means fires are sometimes started. For instance, who would ever suspect that the common or garden rat possessed all the qualities of an incipient fire-bug. In the city of Washington, during one year, 36 outbreaks arose through rats nibbling at the ends of matches; proof sufficient that where fire is concerned not even the most remote possibilities can be overlooked with impunity.

The prevention of panic in schools, shops, factories and the like is, of course, one of the most important features of the ethics of fire-fighting. It is no exaggeration to say that as many people are killed by suffocation, by being trampled to death and by unnecessarily jumping into the streets, as are actually sacrificed to the flames themselves. Human nature is easily susceptible of control, provided there is at hand a sufficiently strong influence to inspire confidence and restore nerve. This influence must be a combination of self possession and training; with this upon which to draw, panic can often be averted. Thus in schools, teachers should be trained in the marshalling of their charges in the same way that employees in shops should be taught to look after the safety of purchasers. The timely playing of the orchestra in a theatre has often prevented disaster, and such aids are worthy of more than passing attention. All this has received careful study in the chapter devoted to the subject and the writer confidently anticipates that if his advice is followed, advice framed upon forty years of actual experience, the casualties due to fire panics will be appreciably minimized.

These are some of the issues connected with fire-fighting, which have been dealt with in as exhaustive and interesting a manner as possible in this volume. The particular intention of the writer has been to avoid lengthy and tedious explanations, which would be beyond the comprehension of the untrained layman. To that end an appendix has been supplied replete with all the tables necessary to the scientific fireman. For the rest, the problems of fire control have emerged from the chrysalis stage of experiment into the fully developed formulæ of an exact science, and the time has arrived when no one can afford to be ignorant of the first principles governing the same. A great quantity of useless information is assimilated by the public; is it too much to hope that opportunity may be found for the perusal of a subject so closely connected with the welfare, safety and homes of the people.

AMBIDILAC

CHAPTER II

•

FROM the earliest times the Romans well recognized the ever present menace of fire and as a matter of precaution a law was passed compelling the erection of separate houses, each standing in its own plot of ground. But as the size of the city increased this regulation became more honored in the breach than in the observance, with the result that serious conflagrations occurred frequently and thus the subject of effective "fire-fighting" was forced upon the attention of the authorities. Indeed, there is nothing surprising in Rome having been constantly visited by such calamities. The houses in the poorer and more populous quarter of the city were usually constructed of wood, sanctuary fires were continually kept burning in every household in honour of the domestic deities and it does not require the imagination of a Jules Verne to conjure up visions of the dire results caused by an act of carelessness or a moment's thoughtlessness. The streets being narrow and tortuous, the smallest blaze would quickly develop into a veritable conflagration, the magnitude of which would depend solely upon the natural barriers which might stand in the way of the flames. In addition, intermingled with the dwelling houses, were vast warehouses and granaries which offered an easy prey to fire.

Furthermore, human nature in Ancient Rome was much the same as human nature in modern New York and enterprising miscreants were not lacking, who realized that by starting a fire and availing themselves of the ensuing confusion, they could enrich themselves comfortably

7

and quickly at the expense of their neighbours. They were, in fact, the germ from which developed the individual who is a terror to his neighbours, a pest in the community and a source of constant activity to fire departments, by whom he is dubbed expressively a "firebug."

Hence it will be seen that even at this early date the menace of fire in its primary conditions did not differ materially from the modern fire risks in many towns. Under the Republic one of the duties of the Roman "Triumvirs" was to protect the city from fire, and later they came to be called "Nocturns," because of their mounting guard during the night. In this task they were assisted by the "Ædiles," to whom the care of the buildings in the town was entrusted. This constituted the "official" fighting force, but there were in addition private organizations consisting of slaves, whose services were given gratuitously according to the wishes of their masters, who doubtless in this manner hoped to rise in public esteem. This forms an interesting analogy to the methods employed by many so-called philanthropists of the present day who are usually ready to support any public work upon which a liberal amount of limelight is turned.

Little could be expected from a department composed of such heterogeneous elements, ignorant alike 'of discipline and organization. The Emperor, Cæsar Augustus, realizing the importance of effective fire protection in his capital, introduced the first regularly constituted "fire department" known to history. It consisted of seven cohorts, each numbering roughly one thousand men. Their duties consisted not only in the actual work of fighting the flames, but also in policing the streets contiguous to an outbreak and in preventing robbery and looting. The fire chief was known as the "Præfectus Vigilum." He was assisted by three lieutenants, "Subpræfecti"; seven "Tribunes," forty-nine "Centurions," and a great number of "Principales." This last title was given to every one in the Roman army, who had any species of fixed office, to all those in fact who occupied the intermediate ranks between commissioned officers and common soldiers. Prominent amongst the "Principales" were the "Librarii," who kept the accounts and paid the wages, the "Bucinatores" or buglers, the ensign bearers, one for each cohort, and the "Aquarii," the "Siphonarii," the "Sebaciarii," and the "Mitularii," to whose respective duties attention will be paid when considering the manner in which fires were fought. There were also four doctors attached to each cohort and last, but by no means least, an official known as the "Questionarius," whose interesting duty it was to apply torture in cases of suspected incendiarism.

The seven cohorts were quartered in as many barracks, designated "castra," which were so located that each could effectively protect two of the fourteen regions into which the city was divided. As to the construction of these barracks, there is fortunately preserved an important record in the shape of a fragment of an ancient plan of Imperial Rome, showing the details of the barrack allocated to the first cohort. This was situated near St. Grisogone in Trastavere and the building had evidently been specially designed for the use of firemen on duty. The atrium or entrance hall was tiled with black and white mosaic arranged to represent various marine subjects, while in the middle stood a handsome hexagonal fountain. Flanking the walls on either side were benches for the men, while numerous inscriptions and rough drawings evidenced the fact that in their moments of leisure the Roman firemen found amusement in caricaturing their fellows. Opposite to the main entrance of the atrium was a door leading to a spacious bathroom, giving the impression that the wants of the men, even in those days, were the subject of as careful consideration as they are today.

It must have been about this time that the intellectual activity of the Romans commenced to assert itself and not only the great "Thermes" or baths were open the whole night long, but also such halls of assembly as the "Palestræ," the "Scholæ," the "Bibliothecæ" and the "Pinocotecæ" would be crowded at all hours with throngs of eager disputants. In fact, nocturnal life in Rome had come to be an integral part of the city's existence.

This in its turn necessitated some form of municipal illumination and this was likewise entrusted to the fire department, a special branch being formed under the name of the "Sebaciarii," after their first captain, one Sebaciarius. Special men were drawn monthly from each cohort for this service, their duties including the supervision of the monster torches kept continually burning outside fire stations as a signal to all and sundry whither to repair in the event of wishing to give an alarm of fire. Some few years ago a bronze torch was excavated not far from St. Grisogone, which experts presume to have been a street lamp of this period.

Fortunately Rome was well supplied with water, which was carried in "Hamæ" or light vases by squads of firemen to the scene of an outbreak, where it was placed at the disposal of those in charge of the "Siphones" or hand pumps. From specimens, which have been frequently found in excavations, these latter must have been very similar to the old-fashioned syringes used by gardeners, only, of course, constructed of wood.

The "Aquarii," or as their name designates, the water carriers, did not confine their attentions to that duty alone. They were also expected to be conversant with all possible sources of water supply in the two regions of the town for which their cohort was responsible.

On the whole, the firemen were well equipped with apparatus including hammers, saws, mattocks, and other such implements, besides leather hose in suitable lengths. Large pillows, specially designed to break the fall of anyone jumping from a height were in general use, and incidentally were not much improved upon till the beginning of the last century. In addition the Roman ladder, the forerunner of the modern escape, had already been introduced and a detailed description of the same may be found in the chapter dealing with appliances.

Given these data it is not difficult to frame in the mind's

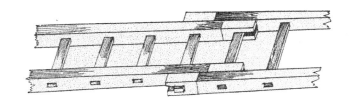

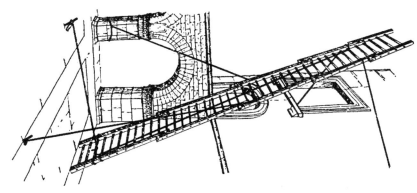

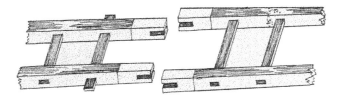

THE ROMAN LADDER. DETAILS OF CONSTRUCTION AND METHOD OF RAISING

eye a picture of a fire in ancient Rome. There is sufficient evidence that the Romans were distinctly human and no doubt an outbreak of fire provided a pleasant interlude when the discourse of a popular orator started to become tedious. Hence it can be imagined, even as today, that the "Nocturns," or fire police, were fully occupied in preventing the curious from hindering the firemen. The "Præfectus Vigilum," or Fire Chief, would arrive to take charge of operations and woe betide any one in the vicinity, were there any suspicion of incendiarism. The services of the "Questionarius," or Fire Marshal, would be hastily requisitioned and, judging by the comprehensive fashion in which the law was administered at that period, it may be hazarded that while no doubt the guilty eventually received their well merited reward, it is not unlikely that meantime a proportion of the innocent had also tasted of that official's ingenious skill. This assuredly must have had a discouraging effect upon the enthusiasm of the genus "firebug," for inasmuch as example is generally a deterrent, it mattered little whether the punishment reached the real offender, so long as the "modus operandi" of the punishment and the reason thereof were known and appreciated.

But to return to a more serious vein of thought, it is a fact that modern methods of procedure against incendiaries lack the finality and thoroughness of those early days. In a later portion of this volume the subject is treated at length and hence it is unnecessary further to pursue the question. Suffice it to say, that, broadly speaking, the Fire Department of Ancient Rome was as well organized and equipped for its duties as many a municipal force as late as the eighteenth century and it might not be exaggeration to hazard, even composed of as competent fire-fighters as some corps of today.

CHAPTER III

IT may be safely asserted that the fire department of ancient Rome was better organized and better equipped than the rough and ready volunteer services maintained by the great European cities during the middle ages. There had, in fact, been a period of retrogression, which was coincident with the dismemberment of the Roman Empire, when all art and science languished in the chaos that ensued. Needless to say, the problems affecting fire control were relegated to the background, and, indeed, the art of destroying towns received more consideration than that of their preservation. Thus it is that no records can be found of mechanical appliances being used at the conflagrations which demolished Constantinople and Vienna. Indeed, this retrograde movement had so far affected the whole subject that even in the Renaissance, when Europe teemed with fresh ideas and new thought, no other method of fighting fires existed than the primitive bucket of the Pre-Roman period. By 1590, however, there were signs of an awakening interest and in an account of a fire in England the use of a monstrous syringe is related as the introduction of a novelty, although in reality it must have been practically a counterpart of the "siphonarius," mention of which was made in the last chapter. In 1615, a hand engine was made in Germany, but it was merely a pump without hose, the principle embodied being a rotary paddle wheel, which by being turned rapidly forced the water out through an orifice. This again was not new, the idea having probably been derived from Greek sources. Even in 1666, the good citizens of

London were without any mechanical appliances and were practically helpless to stem that terrific conflagration, which devastated their city and consumed 13,200 houses covering an area of 436 acres, the ancient cathedral of St. Paul and thirty-six other churches, the Royal Exchange, the Custom House, hospitals and four prisons, in which, incidentally, several persons lost their lives. The value of the property destroyed amounted to nearly sixty million dollars, and it undoubtedly served to impress upon the public mind the necessity of some proper system of fire prevention. Immediately afterwards the city was divided into four districts, each under the control of a special officer possessed of authority to take charge in the event of a fire.

It must be understood that at this time social and economic conditions made life comparatively simple. Gas and matches were unknown thus eliminating those two fruitful sources of carelessness. Buildings were as a rule one-story in height and the floors, even in the dwellings of the wealthy, were flagged with stone. Hence the change was slow in coming and was concomitant with the demand for increased security of persons and property. Business activity began to show itself in all parts of Western Europe in the fifteenth century, and towns destined to be the industrial centres of the modern world had their genesis. With their growth began afresh a full appreciation of fire risks and the necessity of fire control. Yet it was not until the eighteenth century that one Richard Newsham designed a hand engine of practical utility. Water was supplied to it by hand and was then pumped out through a hose, thus forming the predecessor of the manual, draughting its own water and thereby supplying pumps.

America had to learn her lesson in her own way. From the Atlantic to the Pacific her colonists found the country covered with dense forests, which were naturally utilized for building purposes, and, as a result, as early as 1648 the first fire ordinance was adopted in New York, forbidding the use of wooden chimneys and providing for the purchase of one hundred leather buckets, hooks and ladders. A body of volunteers was organized to patrol the streets at night and watch for outbreaks, who from their persistent, painstaking and sometimes rather indiscrect efforts were christened suggestively, "The Prowlers." Their work was, however, appreciated, and in 1678 the town of Boston organized the first regular fire company under municipal control and imported from England a species of hand pump. Only in 1808 did a Philadelphia firm put on the market riveted leather hose, and soon afterwards an ingenious hose carriage of American invention was adopted and remains in use in a modified form to the present day. England was the first country to manufacture rubber hose, about 1820, and its employment with certain improvements has become general. The application of steam as a means of obtaining power was responsible for a revolution in fire apparatus as it was in all other lines of mechanical effort. It has contributed in no small degree to the construction of effective portable machinery with which to fight fires, and the benefits derived from its use have been almost incalculable.

Obviously, it is the endeavour of all firemen to check a fire in its early stage, since generally speaking, its commencement is small and progress comparatively slow. It is no exaggeration to say that some of the great conflagrations which for hours and even days have baffled the combined efforts of huge fire departments with scores of determined firemen equipped with much powerful apparatus, could have been extinguished in a few seconds by the cool-headed and well-directed work of one man armed with but a single pail of water, had he arrived in time. In other words, if ready means of suppressing a fire in its infancy were at hand many serious outbreaks might be averted, and hence it is that so much depends upon effective apparatus and the speed with which it is conveyed to the scene of action.

For imagine what happened in the old days before the adoption of the steam fire engine. First, consider the

bucket period. A person discovering a fire would run to his nearest neighbour for help, and then the alarm would be given from one house to another and immediately all would be confusion. Volunteers there would be in plenty, armed with buckets or any other domestic utensil which would contain water. Forming a line they would pass the buckets from hand to hand, sending them back by their women folk to be refilled. With such loss of time and feeble resistance it is small wonder if usually the flames continued their course practically unchecked, and a building saved from complete ruin was considered as a remarkable achievement.

Next came the period of the hand engine. Bells upon churches and other public buildings were now the means of spreading the dire tidings, and upon hearing their summons the voluntary firemen would hurry to their quarters and drag their engine in the direction of the first alarm. Then arose the question, where was the nearest water supply, and no doubt time was wasted through unsolicited advice. If, as was often the case, the supply proved to be at too great a distance from the outbreak for one engine to furnish an efficient stream, a second was stationed between the fire and the water. The ensuing contest between both parties of excited men as to which should occupy the place of honour near the fire, and the efforts of the vanquished to pump up more water than the engine in front could use, no doubt, added to the gayety of the community, and the mythological God of Fire must have smiled and perhaps murmured, "What fools these mortals be." But this opera bouffe method of fire-fighting really served a useful purpose, inasmuch as the increasing seriousness of the fire risk did not appeal in the same degree to the sense of humour of those who lost their property, with the result that the advent of a new factor in fire control was welcomed by the influential of the population. George Braithwaite, an Englishman, first conceived a steam fire engine, which was completed in the year 1829 and was a portent of the great change to come. Skeptics there were who scoffed at its superiority and who jeeringly referred to it as the "steam squirt" or the "kitchen stove." But it had come to stay, and in 1840, a New Yorker, by name Paul Hodges, constructed a model of curious design, which, however, proved impracticable.

The year 1845 was marked by the first of the great fires, which heralded the era of new building construction in the United States and which, therefore, deserves more than passing mention. Pittsburgh, Pa., was the scene of the disaster, which originated from the simplest act of carelessness. On washing day, in the early part of April, a house-wife made a small fire, upon which to boil water, in the back yard of her home. A high wind was blowing and sparks from her miniature bonfire were carried to a neighboring building, which quickly ignited. With incredible rapidity the flames spread from house to house, and, despite the desperate efforts of volunteer and amateur firemen, the destruction ceased only when no material was left for the fire to feed upon in a territory fifty-six acres in extent. The financial loss was five million dollars, an enormous one for those days, and two thousand families wandered homeless over the charred remains of what had been their dwellings. This is one of those instances when prompt and timely action would have probably saved the situation, but the antiquated methods employed, coupled with the delay inseparable from the summoning of volunteers, was just sufficient to transform what might have been a back-yard blaze into a conflagration of the first magnitude. And so it always will be in fire control; time is an ally of the utmost value, which in its turn demands the maximum of celerity on the part of all concerned. Prominence is given to this episode since some such reminder was needed in America, as elsewhere, to stir up its citizens to a realization of what fire could accomplish, even from the smallest and most trifling beginnings.

Untold romance lies in the history of the great forest fires of America, which even today rage to a large extent uncontrolled, but which educated the early settlers to a vivid realization of their perils. Thus, in the prosperous colony of New Brunswick there is chronicled a conflagration, which in its destructive horror has left an indelible mark upon the population as well as upon the land itself.

Along the banks of the river Miramichi there were scattered in 1825 prosperous settlements of fishermen and farmers, while through the forest which extended for hundreds of miles to the north and south roamed hunters and trappers of nomadic habits in search of a livelihood. To them, nature appeared so bountiful, that no thought of any enemy common alike to both entered their contented minds. The summer had been a dry one and the autumn had brought but little rain, till the pine needles and leaves crackled under the weight of a passing step. And a careless lumberman was to transform this haven of quiet into a holocaust of ruin! Having finished his evening pipe, he knocked it out against a tree stump and turned in, little recking of the consequences. He awoke to find the forest ablaze about him, and although fearfully burned managed to make his way to the nearest camp, where there was no need to tell his story. For east and west, north and south the glow of an unnatural day was upon them. From the waters of the Miramichi to the shores of Bay Chaleur, there was one roaring hurricane of flame and no human means wherewith to stay it. Dawn followed dawn bringing no relief, till the heart of a great province was transformed from a richly wooded country into a lonely and desolate waste. So much had been accomplished by human carelessness, though it is ever thus that the world has learned its lessons. No less than two hundred persons either perished in the flames or were drowned in the river, vainly trying to find safety in its cooling waters. Over a thousand horses and cattle were swept to their doom, and six thousand square miles of forest disappeared as completely from the face of the earth as though they had never been. In some places the destruction of vegetation was so thorough that even to this day nothing can grow there but stunted shrubbery and coarse grass, a constant reminder of this tragedy. With such examples of the terrific power of fire, was it surprising that the new world hailed the invention of the steam engine with enthusiasm as a possible panacea for its sufferings?

Even to the amateur mechanic the principles governing the construction and working of the steam fire engine are simple and easy of understanding. In the earliest examples an upright boiler with a spacious fire box at its base was set between the rear wheels of an ordinary carriage body, and surmounted by a short smoke-stack. Bolted to the front of the boiler were two steam cylinders, above them being placed the pump itself, so that the piston rod of the engine served as the rod of the pump. Steam drove the pistons up and down in the engine, drawing water through a large suction hose on one side and forcing it out on the other through a smaller hose. From the pumps the water was forced to an air-chamber forming a cushion and serving to equalize the pressure, thus giving an unvarying discharge. The principle of these pumps was, therefore, very much akin to that of the hand engine, but with enormously increased power. As this was long prior to the introduction of the water tube boiler, steam had to be generated in the old way, by which the heat given off by combustion is conveyed by tubes through the boiler. The water supply of the boiler was obtained from a small pipe connected to or near the suction chamber and pumps. On the average the diameter of the cylinders in the various sizes of engines ranged from six and one-half to ten inches, while the stroke as a rule measured eight inches. These rough particulars will give the reader some idea of the chrysalis from which the modern fire engine has emerged.

Since fires cannot be fought without water, some account of the problem connected with its supply deserves attention. Here again may be observed the retrograde movement, since in Roman times it was not uncommon to find aqueducts forty miles in length, which, from their situation, were enabled to deliver to the city, in accordance with the laws of gravity, a sufficient quantity of water

CANTINIÈRE OF A FRENCH RURAL FIRE BRIGADE.

at a moderate pressure. Naturally this was of great advantage in fire-fighting and from historical records it is clear that the most was made of it. But in Europe of the middle ages, these lessons had been forgotten and the practice had fallen into desuetude. Rivers, wells and ponds were considered adequate for the needs of the population and it is curious to meditate that the intellectual wealth of that time expended itself solely upon art and the most profound metaphysics to the exclusion of more mundane, though probably more useful, considerations regarding public health and safety. Yet, even in the middle of the last century, it was by no means uncommon to find large towns dependent upon a water supply operated by private companies and conveyed by means of open mains through the streets. In 1815 Philadelphia introduced a complete system of underground pipes constantly supplied with water by a central pumping station. This plan proved a success and has since been gradually adopted even by many of the smallest towns in America. This system, however, did not at its initiation take into consideration the fire department, and the city of New York probably had the first water service to which hydrants were connected for that particular purpose. By degrees has been evolved from this mode of supply, that most valuable adjunct of modern fire-fighting, the "high pressure" system, which even now has not been extended to its limit of usefulness and which is lacking in cities where it should most certainly be installed. A detailed description of its advantages is given in a separate chapter.

Naturally, an outbreak of fire being invariably attended with some danger to human life, those far-sighted Romans cast about for the most simple yet effective means of coping with the situation. Two pieces of their apparatus were specially designed for this purpose and have survived in a modified form until the present day. Firstly, mention must be made of the Roman ladder. The great advantage of this apparatus lies in its simplicity. In its constructive details it has changed practically not a whit since the days of Nero, and it is as useful in wide thoroughfares as in narrow courts, while its portability is such that one man can carry the entire equipment. It consists of a series of short ladders from six to nine feet in length, the lower part of each being slightly broader than the top. By means of a slot the sections can be fitted together, all being interchangeable except that designed for the bottom, which has its sides somewhat more outspread in order to provide a firmer hold upon the ground. The method of erection is simple and ingenious. The lowest section is first placed against the wall to be scaled, at a considerable angle. The fireman then ascends it with a "section" on his shoulder and armed with a rope, a hook belted to his waist and a pulley. When he reaches a certain rung, which in modern practice is painted scarlet, he puts his leg through the ladder, his foot against the wall, and hooking himself on, in order to leave his hands free, pushes the ladder away from the wall and fits the "section" he has carried on the top of the "section" upon which he is standing. He then hauls up another "section" and repeats the same manœuvre. At the Colosseum in Rome, for exhibition purposes, these ladders have been joined up together till they reached a total length of one hundred and sixty-four feet. This apparatus, it may be remarked, is in regular use in many of the Italian fire departments today.

The second noteworthy appliance of Roman times, which has endured through all these centuries, and which in the writer's humble opinion, modern invention has not improved, is the jumping pillow. This was nothing more nor less than a large mattress some eight feet square, stuffed with hair or feathers, and designed to break the fall of any one jumping from a height. Nowadays the practice is to use a net made of heavy rope attached to springs to afford additional resiliency. The chances of any one jumping from a height of more than three stories must always be intensely hazardous, but all things considered, there appears to be a balance in favor of landing on the pillow. During that most distressing fire at the

Asch Building in New York, when a number of lives were lost, several young women attempted to jump to safety, were caught in the net and found—death. The impetus their bodies gathered while falling was so terrific that the shock of the impact killed them in every case. Hence it will be seen that the fire-fighting world is still awaiting the genius of the inventor, who will be able to devise some other means of catching unfortunates who are compelled by dire necessity to jump to their doom.

This brief résumé will have been sufficient to demonstrate the fact that the inclusion of fire-fighting amongst the scientific problems of the day and as one worthy of serious consideration dates from modern times, and hence the many improvements which have been introduced into its practice are all of such recent origin, that even now they are only just emerging from an embryonic stage. It is probable that the next century will witness advances along all lines, of such immense consequence that present apparatus will be totally outclassed and will be relegated to the glass cases and dusty environment of museums, where the curious of future generations will gaze with interest, tinged possibly with amusement, at the appliances used to fight the flames by their forefathers. So far the use of chemistry as an ally of water in subduing fires is only in its infancy and though prophecy is admittedly unsatisfactory and more often than not misleading, it may be hazarded that the cumbersome steam fire pump will in due course disappear from the sphere of active operations and that the outbreaks of the future will be dealt with swiftly and easily by a combination of high pressure streams coupled with chemical forces as yet inoperative. It has taken many centuries to evolve the fire departments of the present, but as so often happens, now that a scientific advance has at last been made, that advance will continue with increasing rapidity until "fire," as was always intended, shall be the "servant" and not the "scourge" of man.

CHAPTER IV

A SAGE once penned the dictum that "Fire makes a good servant but a bad master," and few practical firemen will be found to argue the accuracy of the statement. For the fire-fighter, life consists of one protracted struggle against this most crafty of elements, which oftentimes is most dangerous when apparently subdued and which, in its methods of attack, would appear to be guided by some Machiavellian master mind of strategy. Hence it goes without saying that successfully to cope with such an adversary demands the maximum of skill and determination, which are fundamental characteristics of the genus fireman. The sailor is an idol of the public largely because he is ever pictured as pitting his seamanship and science against the two stubborn forces of wind and waves. In song and story he is immortalized as the acme of all that is dashing and fearless, and it is small wonder that the younger generation, inspired by such narratives, yearns to emulate such heroes. Yet, for some strange reason, the fireman has never occupied so large a place in popular romance; his deeds have not been chronicled with the same degree of graphic narration, the cheap notoriety of the music hall ballad has perhaps happily been denied him, and it has remained for the daily press to utilize him as a convenient "feature," in the absence of other material. This must not be taken as implying any want of generosity on the part of those concerned; but naturally a minor fire, though involving considerable risk to those operating against it, cannot receive the same publicity as that accorded to some event of general interest. Also, it must be remembered that it is a common trait of human nature to accept without particular comment the services of any organization to which it has become accustomed. The average person is ignorant of the sea, except through the medium of what is written, and hence, being unfamiliar otherwise with the subject, instinctively envelops the calling of the sailor with a glamour of romanticism and mystery.

Nevertheless, to those who care to seek it, there is a potent fascination in the career of a fireman; a life full of ever-varying incident and a calling which may well make appeal to the imagination of the young man in search of adventure.

Picture a warm, well-lighted recreation room. A dozen firemen are gathered about tables passing away the time with dominoes and pool, while one of their number is amusing himself at the piano by strumming over the latest popular airs. Suddenly the alarm gong sounds. It is a district call, and almost instantaneously the men are in their jerseys and boots. The pianist has disappeared down the sliding pole with a celerity which would put to shame the demon in a fairy play. While the others are following, the horses have already clattered from their stalls and taken their allotted places under their harness. With a snap and a click their collars are locked, the drivers leap to their seats and as the station doors swing automatically open, the firemen clamber onto the apparatus, which is already under way. "Not so bad," mutters the officer in charge, "eighteen seconds from the alarm." Outside the night is chill and misty, intensified by a steady drizzle. The streets are greasy and the engine rocks perilously from side to side as, with bell clanging and siren sounding, it dashes at full speed along thoroughfares crowded with home-going pleasure-seekers. Arrived at the scene of the outbreak it is found that the cause of the trouble is a large warehouse on the waterfront, full of combustible materials already well ablaze. The driver, versed in the geography of the district, pulls up at the nearest hydrant. A loud clattering heralds the approach of the hose wagon. A burly fireman deftly catches the end of a hose section thrown to him and couples up to the standpipe. Then the crew, with an automatic precision born of long experience, lay hold of their weapons, the hose pipes, and advance to the attack. Above the roar of the flames raging up the elevator shaft of the building resounds the shrill crescendo of a ship's whistle. A fire boat has responded to the call and is wending its way rapidly along the water's edge. Within a period, measurable only by seconds, it also has joined in the fray and is directing several streams upon the rear of the main fire, thus carrying out the most effective manœuvre in modern warfare, that of outflanking the enemy. Meantime other engines and other apparatus have arrived. Curious crowds have collected and strong drafts of police are kept busy in preventing the hampering of the brigade's efforts. A large motor draws up. Its occupant is the fire chief, distinguishable from all others by a white helmet. There is no confusion, little excitement, the General has arrived to take the supreme command. An officer briefly outlines the situation; the fire has gained such and such a hold, so many pieces of apparatus are being employed with a certain end in view, the only question is whether the General is satisfied that the forces are being used to the best possible advantage. They decide that a personal inspection is necessary and without delay the Chief enters the building. Nearby stands a hospital ambulance, with its doctor and orderlies ready for any emergency, for even as on a battlefield casualties are to be expected. By order of the Chief a heavier attack is developed upon a particular portion of the structure and an extension ladder shoots up through the murk with men clinging cat-like to its rungs even as it lengthens. An order rings out "Start water," and a powerful jet is forced into the heart of the seething inferno.

The crucial point in the attack upon the fire has now arrived. It is as though each contestant were summoning up his reserves with a view to one overwhelming effort at

"FIRST AID" HYDRANT AND
HOOK AND LADDER CART OF LUCERNE.

mastery. Flames have crept into the cellars, rendering the task of the firemen in that quarter almost impossible. Several are overcome by heat and smoke and are quickly removed to the ambulance, their places being speedily taken by reliefs. But still the fire gains. Moreover, a new ally assists the flames in the shape of a snapped, heavily-charged electric light cable. Like some huge serpent it twists and writhes hither and thither, menacing with instant death those, who again and again essay to check its career. It hisses venomously, its blue glare blinds them in the pervading gloom until with one supreme effort it is seized and denuded of its fangs, being severed from the main. One successful skirmish does not, however, constitute a victory, and a reinforcement of the enemy appears to check too confident an advance. The roof is yet intact and upon the third floor the firemen are met by great volumes of dense smoke, which threaten "back draught." With axes and hatchets, doors and shutters are demolished, anything to create a draught. A sheet of flame and a whirling eddy of sparks momentarily envelop the workers on the extension ladder and few among the watchers can credit their safety as they emerge from this fiery whirlpool, clothes burnt, hair singed, hands blistered, but still fighting on with grim determination. That marked the last desperate stand of the enemy. The Niagara of water is beginning to tell and a sullen pall of smoke darkens the angry brilliance of the blaze. Some of the companies are recalled to their stations to be in readiness for other outbreaks, while a sufficient number of men remain until the last vestiges of their foe have disappeared. Then they, too, retire, perchance to a well-earned rest.

This is by no means an over-coloured picture of an everyday fire in the warehouse district of any city; moreover it is devoid of the heartrending scenes and nerve racking uncertainty inseparable from those occasions when human lives are involved. Thus, who shall say that the life of the fireman lacks that romance, which is supposed to be inalienable from them "that go down to the sea in ships, that

do business in great waters." As a matter of fact the career
of those who fight the flames teems with anecdotes of splendid courage, self-denying heroism and hair-breadth escapes,
which furnish material and to spare for the great masters
of the pen. For instance, this from real life:

During the progress of a serious fire in the city of Boston,
the Assistant Chief went to the top of the building involved
for the purpose of opening the hydrants connected with
its water protection. Whilst thus engaged he was cut off
from all means of escape save one, which consisted of a
heavy telegraph cable connected to a separate building
across the street. In order to make his predicament known,
he threw his fire helmet to the ground, many feet below.
Extension ladders were erected with all rapidity, but were
prevented from reaching him by a tangle of overhead wires.
By this time his clothing was on fire and the position was
rapidly becoming untenable. All that separated possible
life from a horrible death was that cable. Crawling to
the edge of the building he swung himself onto the wire,
which swayed and quivered with his weight. With the
utmost presence of mind, hand over hand and leg over leg,
he worked his way toward the centre of the cable, where
he remained suspended ninety feet above the ground. Had
the line run directly across the street, the officer, with the
distance he had actually covered, would have reached
safety.

But unfortunately the line was at an upward angle and
his efforts to reach the point which he had gained had
sapped his vital energy to that degree which made further
progress impossible. Men were hastily placed on hose
wagons, which were backed in together to form a circle. A
life net was then stretched between them, in case his strength
should give out and his grasp relax. Fortunately at this
juncture, one of the firemen with a special knowledge of
knots made his way to the roof of the house upon which
the fire-free end of the cable was attached. Fastening a
rope to the cable, he sawed the latter through, thus enabling
both man and cable to be lowered inch by inch towards

the ground.   When the knot joining cable to rope was reached, the officer lost his hold and was caught by his comrades in the net and carried into the street.   This exciting escape proved no barrier to the further duty of this fireman, who twenty-four hours after this incident was able to report for service and "carry on" as though nothing had happened.

In the early days before fire departments had come to be officially recognized, dependence had necessarily to be placed upon volunteers and many are the stories, humorous and pathetic, which could be told about them.   The fascination of the service certainly extended to those who enrolled, judging by their social position and by the fact that many of them gave up valuable time in order the better to qualify for their duties.   Some peculiar entries are to be found in the old minute books of these stations, indicative of the fact that the commonest breach of discipline would appear to have been a too free use of strong language.   Thus, the secretary of one company reports a fireman for saying to him, "You be damned, you damned old Dutch hog," for which he was severely reprimanded; while the puritan spirit was carried so far that a man was fined for saying "Damn the odds."

Some fifty years ago it was customary for all young men to belong to associations of some sort, religious, social, or political.   The story goes that one such youth was sitting in a tavern and overheard others of his age discussing the societies to which they subscribed.   This filled him with a desire to go and do likewise, so on his return home he told his mother of his ambition, remarking that he was not particular as to the nature of the club which he joined.   There was a great revival going on in those days and like all good mothers she told him to go with her and join the church.   "Well," quoth he, "I don't specially care what it is, but I must belong to something."   So down to the church he went, but to his chagrin the minister told him that he must be placed on probation for three months.   When that period had expired he was told that he must wait yet another two months. Some time passed, when one day the minister happened to meet his probationer walking down the street in a neat, red shirt, a gaudy pair of suspenders, a coat thrown over his arm and bearing a number on his back. "Aha," said the pastor, "you're the one I want to see. You haven't been to church of late." "No, Dominic," answered the young man, "that probation was too long for me." "But," cried the former, "it is at an end and you can now join the church." "Too late, too late, Dominie. I've joined an engine company down here and it's going to take all my time to look after fires. I'm going to one now. You see I was bound to join something and these fellows let me in without any probation; all I had to do was to give up two dollars and I was called a member. Come round to see us Dominie, we've got as bully a little engine as ever went to a fire." From which it may deduced that the pleasures of earthly fires were greater to the majority of young folk than the terrors of the fires to come, as depicted no doubt in the Bible meetings.

About that time one of the most popular chiefs was James Gulick, who commanded the New York Fire Department. The following incident is illustrative of the affection in which he was held by his men. A fire had broken out in Centre Street, adjoining the works of the New York Gas Company, which had destroyed two houses. Against the gable end of one of the burning buildings a large number of barrels of resin were piled, and the firemen worked diligently to save them by rolling them into the street. The night was intensely cold, and somebody kindled a small fire with a part of the contents of a broken barrel, which the workmen employed by the gas company attempted to extinguish. These were warned by the firemen to desist, and a big, heavy fellow who continued his efforts was pushed away. Thereupon a large number of his friends attacked the few firemen in charge, who were joined by their comrades and a fight ensued. The brigade was victorious. Gulick heard of the affair and hastened to the scene exclaiming, "What does all this shameful conduct mean at such a moment." The only answer was a blow from a workman, who struck his head from behind with an iron bar, and only his helmet protected him from serious injury. Turning upon his assailant the Chief pursued him across the ruins of a fallen wall and threw him upon the débris, but was followed in his turn by some thirty or forty employees. "Men, stand by your Chief," was the cry of the devoted brigade, and in an instant the attack was turned into a rout, the workmen taking refuge in the gas house. Gulick, by almost superhuman efforts, forced an entrance in advance of his enraged followers, and amid volleys of coal buckets, called upon the rioters to surrender, promising protection. His reply was a charge with a red-hot poker, which fortunately passed through his trumpet, which he carried under his arm. This put an end to his forbearance and jumping from the doorway, he shouted, "Now men, surround the house; don't let one of them escape." The excited firemen rushed into the building and administered a sound thrashing to their truculent foes, who were afterwards arrested, and even then the former were not appeased and attempted to destroy the machinery which was only saved by the Chief's firmness and discipline.

After the great fire of 1835, which caused twenty million dollars' worth of damage and dislodged more than six hundred mercantile firms, the resignation of Gulick was de manded, upon which the brigade *in toto* struck work and it was only with the greatest difficulty that it was reëstab lished on a satisfactory basis.

Perhaps the writer may be forgiven for trespassing upon the patience of his readers to the extent of drawing from his own personal experience some anecdotes illustrative of the various phases of his life, both before and since he became a fireman. If there is any truth in the old adage that *experientia docet,* then assuredly thirty years of practical fire-fighting in the largest organization of the kind in the world entitle him to form some opinions and arrive at some conclusions. It would not be difficult to write a whole book with the personal material at hand, but the present object is rather briefly to show how any young man minus influence or capital, but possessed of determination, may climb the ladder leading to positions of grave responsibility and ultimately to the head of his chosen profession. Incidentally the writer wishes to emphasize the fact that his advancement was in no way due to any exceptional opportunities or to what is termed popularly good luck, but rather to a steady and unremitting attention to duty, coupled with some of that perseverance which in that historic race between the hare and the tortoise gave the victory to the latter.

Since the following narratives are the writer's own experiences, it seems more apropos to relate them in the first person.

At the immature age of three I may claim to have received my baptism of fire, since like most other youngsters anything to do with the forbidden joys of matches possessed an unholy fascination for me. One day while playing with some other children, whose tastes were similar to my own, I conceived the brilliant idea of making a good blaze in the hay yard. I cannot remember whence I procured the matches, which in those days were a great luxury and were carefully hoarded, but since desire is the father of acquisition, by hook or crook I secured some, and what could make a better bonfire than a haystack? Within less time than it takes to write, one was in flames and we jumped and danced around it playing at Red Indians, until some unsympathetic neighbours came running from all directions gesticulating wildly. It then occurred to me for the first time that I had done wrong, and I promptly showed a clean pair of heels to avenging justice. Running into the house I hid under the bed and while workmen and friends busied themselves in saving the house, I lay there not daring to emerge. Not until the excitement had subsided were enquiries made as to the origin of the fire and knowing my foibles I, of course, was suspected and

a search was instituted for the incipient firebug. It did not take long to discover me and drag me forth, when my angry mother carried me to an adjacent stream, telling me that such naughty boys had better be drowned early in life than be allowed to live to burn up property and people. My feelings of remorse can easily be imagined, and I promised that never in my life would I again start a fire and that always I would do whatever lay in my power to extinguish conflagrations. •

But this childish prank, aside from the promise that I made on that occasion which I have ever kept, taught me one great lesson. It is, that children, when frightened by fire, have a tendency to conceal themselves under beds, and, therefore, in searching a dwelling, firemen never neglect to look carefully in these hiding places. When children are awakened by suffocating smoke or by members of the household during excitement consequent upon fire, unless watched they will invariably crawl under beds, thinking in their childish fancy that thereby they are safely hidden from the flames, and many a little body is on that account brought forth lifeless.

It is, of course, difficult to lay down any hard and fast rule for occasions of this sort, but it might be impressed upon children from their very earliest years that under no circumstances should they adopt this method of hiding. Whether in games, or to avoid Mamma with a slipper, the practice is a bad one and though the actual occasion may never arise to prove the value of this instruction, it will undoubtedly, in that odd chance of five hundred, be the means of preserving a precious life. In fact, this is the epitome of fire control, watch and be prepared for the odd chance, for just as the individual who is foolish enough to carry a revolver will probably never need it, but if he does, will need it uncommonly badly, so in all fire precautions, necessity for their use may never arise, but should the unforeseen happen, their absence may prove disastrous.

From my childhood I always possessed a great love for the sea, and thus it happened that at the age of 13, I shipped as "boy" in a top-sail schooner bound from White-haven to Dundalk with a cargo of coal. Her name was the Gazelle, and, judging by her behaviour on that eventful trip, her owners were not mistaken in thus christening her. We left Whitehaven in the middle of an unusually stormy December and by the time we were off the Isle of Man we were running into a howling southeasterly gale, which was not improved by incessant squalls of blinding sleet. Needless to say, I experienced the additional discomfort of being horribly sea-sick, not that on that account I was per-mitted to escape my share of the ship's work. I can re-member, as though it were yesterday, making my way along the wave swept decks and wondering what on earth had ever induced me to leave the comforts of terra firma for such an inferno of physical torment as was apparently offered by a sea life. After hours of incessant tacking we managed to make Belfast Lough, where we found shelter, and anchored preparatory to riding the storm out. The ship was in a terribly battered condition, sails blown to ribbons, boats washed away and half the bulwarks gone. "Ship's boy" in those days was synonymous with "maid of all work," and as there is, so it is affirmed, no rest for the wicked, I was promptly told off to make up a good fire in the "bogie," a dirty little black stove which smoked incessantly and had been the bane of my existence during the voyage. Full of anxiety to disprove the reputation which I had gained as a sea-sick land lubber, I stoked up and soon had a warm, comfortable glow in the fo'castle. Then I turned in.

It must have been half an hour later that I awakened to find the heat becoming oppressive. The cause was not far to seek, the boat was afire.

The black bogie had again played me a low trick and had become red hot. Morever, the flue had caught the infection and, in its turn, was transmitting the disease so effectually that bulkhead and deck planking were emitting a miniature Vesuvius of smoke and sparks. Without waiting for any instructions, I attacked the invader with buckets of water, the sleepy crew lending an extraordinarily willing hand when they realized that their belongings were in peril. On the painful events following the captain's appearance, I will not dwell. Suffice it to say that I received the smartest lacing the old man could give me, the memory of which remained with me long after I had left the merchant service. But the moral is obvious. Anything more ludicrous than stove pipes passing through unprotected wooden bulkheads it would be hard to imagine, yet such is the conservatism of the sea that it is by no means uncommon to find such conditions even today in small coasters and smacks.

The foregoing was my first fire at sea, but I was fated to have another experience of a more serious character. I happened to be quartermaster in the old "Abyssinia" of the now defunct Guion Line, plying between New York and Liverpool. We had sailed from the former port in the month of July, with nine hundred passengers of all classes and a full cargo of cotton. About 280 miles east of Cape Race a fire was discovered in the main hold, which, though located in the middle of the night, was kept from the passengers' knowledge till the noon of the following day, when the united efforts of the crew had been found insufficient to cope with the outbreak. The captain then decided to call upon the passengers to lend a hand, and men and women from saloon, intermediate and steerage, bravely combined with the sailors in their dangerous task. Happily the sea was smooth and, to the lasting credit of all concerned, there was no panic.

Steam was used to fight the burning cotton, and as the seamen were overcome by smoke in the darkness of the hold, volunteers took their places, with the result that after three days of incessant labour the outbreak was under control. Had there been a panic or had the flames gained the upper hand, the result would have been hideous beyond words since there were only boats to accommodate three hundred persons. It only remains to add that Queenstown was made in safety without any casualty, and though the incident lacks any spectacular element, it contains material for thought regarding the principles governing fire control at sea.

The use of steam on shipboard for the extinction of fires is general, though its efficiency is open to serious question. When water becomes steam it is practically non-absorbent, since in assuming this form it has been subjected to great heat. As the object desired in fighting a fire is the absorption of the heat created by the flames, it is apparent that any element at a high temperature is unable to obtain with certainty its reduction. All that can be expected from steam is that by its moisture it may be able to check a further advance of the enemy. Hence, if steam must be used let it contain as much moisture as possible or, in simple language, let it be used at as low a temperature as is compatible with its existence. But in the opinion of the writer the whole subject is one of such a highly complex character, and withal of such overwhelming importance that it merits the study and consideration of all concerned in the safety of passengers and cargo in ocean going vessels.

About the autumn of 1878 I shipped as first officer on a steamer bound from Chicago to Buffalo with a cargo of oats. All went well until we were in Lake Erie, about sixty miles from Buffalo. I had a trick at the wheel from 8 to 12 in the first night watch and on being relieved I went forward to the deck-house, filled my pipe and prepared to enjoy a smoke. Scarcely had I got it well alight when I heard a cry of fire, and rushing out, saw flames bursting through the after hatch close to the companion-way leading to the cabin. The captain, who had been on deck most of the time during the first watch, had gone below a few minutes before. His wife, who was with us on the trip, was in the cabin at the moment. Running aft I realized we had a very dangerous fire with which to contend. The deck watch, in charge of the mate, attacked the blaze and I dashed into the cabin to notify the captain and his wife.

In a few minutes they were both on deck and the fire

had so increased that I suggested the advisability of getting out the boat and launching, in addition, the life raft which we carried. This was agreed upon, since the steamer was constructed of wood and her condition was hopeless. We succeeded in lowering the raft, but the flames had spread with such rapidity that they had enveloped that part of the ship from which the life-boat swung, making its launching an impossibility. Wrapping a blanket around the captain's wife, who was clad only in her night-dress, we were able to get her on the raft, but she suddenly remembered that her jewellery had been left behind and implored her husband to secure it for her. His complaisance almost cost him his life, for on his return to the cabin he was severely burnt about the head and face and he failed, in addition, to gain his object. The dry oats proved excellent fuel, and it speedily became evident that the ship was doomed. We had either to remain by it or to take to the raft, which was built to carry ten persons, while we were fourteen all told. The stokers, engineers and deck hands joined the terrified woman, while the captain, the mate and I went forward to that part of the vessel which was not yet involved in the general conflagration. We stood together near the bow watching the fire advance slowly toward us. The heat was intense and the lake was lighted up for miles around by the flames. Suddenly the fore-mast fell. It barely missed the captain, who stood in a dazed condition by my side. The mate and I realized that in a few minutes we should be forced to jump overboard and made ready by removing our clothing, until we stood only in our undershirts and trousers. From the raft, which was about 250 feet to windward of the burning vessel, came an imploring cry, beseeching the captain to leave his ship and come to his wife. He shook hands with us and sprang overboard. As he was a powerful swimmer, he was soon alongside the raft. We, however, remained where we were for perhaps ten minutes, when it became a question of death by fire or taking our chances in the water. The water seemed inviting in com-

parison with the flames, and we did not hesitate to plunge
overboard after saying "Good-bye" and murmuring a few
words of prayer.   Never shall I forget my sensations when
I felt the cold waters of Lake Erie that October morning!
Actually blistering from the heat, I thought I had been
suddenly transported to Paradise.   Between the pleasures
of dying by drowning and the horrors of being roasted
to death there is a gulf almost as wide as that which
divides the celestial realms from the regions of the damned,
and the sense of security and relief from pain was almost
indescribable.   But now a new difficulty confronted me.

I had learned to swim in salt water and I found the fresh
water exceedingly light and hard in which to keep afloat.
By easy strokes I contrived to get near the raft, but alas,
there was no room for me upon it and any such attempt on
my part would have spelled disaster and probable death to
all concerned.   Floating and swimming by turns I kept up
for about an hour when my strength began to waver and
semi-unconsciousness supervened.   Amongst the crew was
a negro cook who sang songs and cracked jokes in an
effort to keep up the courage of his unfortunate comrades.
All the time that I had been swimming by the raft, this
cheerful creature had watched me, and as I was about to
sink I felt his hand take hold of my shirt and heard his
voice in words of comfort.   He quietly drew me toward
him and with the help of the chief engineer got me securely seated on the raft.   Then he slipped overboard,
where he lay on his back and floated like a chip.   For
seven hours he stayed in the water, helping the captain
and mate alternately to rest on the raft when they became exhausted.   The chief engineer and another took
turns in swimming, but neither stayed in the water as long
as did this sturdy coloured man.   Never once did he com-
plain.   He was the same cheerful soul at the end of his
long trial as he had been when he left Chicago.   We were
rescued eventually by a tug which had put out from Buffalo, having seen the flames sixty miles away.   The memory of that brave negro has always remained with me.

I may say that I owe my life to him, for, though a fair swimmer, I could never have lasted through those terrible eight hours without his unselfish assistance. There has always been in my heart a feeling of gratitude, not alone to the brave fellow who, I am sorry to say, lost his life afterwards in a railway accident, but to the race to which he belonged.

Many years afterwards, when an engineer of a certain company, I had an opportunity of vicariously paying off something of this debt. We responded to a fire, which proved to be in a tenement occupied by coloured people. The building was already a mass of flames and several persons on the upper floors were cut off from escape. Two coloured women and a little boy were trapped on the third floor. Mounting to the windows by extension ladders, we could see them with their clothing already on fire. The only chance of saving them was a desperate one, but we took it. Fireman Malavey and I entered and succeeded in passing the three to others outside, who carried them safely to the ground. The boy and young woman are alive today, but the elder woman was so badly burnt that she died in the hospital on the following morning. It only remains to be said that the one life lost in the Lake Erie fire was that of the captain's wife, who succumbed shortly afterwards from exposure, a circumstance made doubly sad from the fact that she was a beautiful bride of only four months.

Curiously enough, my first active service in the New York Fire Department was in connection with a vessel on fire and is illustrative of the adage that all knowledge is valuable.

As is usually the case with a new member of the force, I was extremely nervous during my first nights at the station. Although my sea-faring life had taught me to be accustomed to turning out any moment and in all sorts of weather, I speedily found that the watching and waiting for the alarm gong possessed a mental strain of its own, which, incidentally, is common to all fire-fighters.

During the night in question I had lain awake with tense nerves, fearing that the call might come and that I might get left behind. Then I fell into a troubled sleep, to be roused by the sound I had so long expected. In my anxiety I stumbled over my own boots and narrowly escaped up-setting my neighbour, who did not appreciate the atten-tion. I gained my object, however, and my nightmare of missing my first alarm dissolved as we galloped through the silent streets.

A French ship was involved, a fire having broken out in the forward hold. With enthusiasm, I seized a length of hose, only to be told in official phraseology to leave it alone. Not comprehending the order, I attempted to board the vessel, but was stopped by the battalion chief, who recognized in me a recruit. Perhaps I may here remark that it took me a full month to master the regular words of command, which are peculiar to fire departments. Eventually I found my chance, for with my marine knowl-edge I knew how best to tackle the trouble, and, creeping along through the smoke, made my way to the heart of the outbreak. There I was found later by the Chief, who, finding me on my face using the hose to the best of my ability, told me to get up, and lending me a help-ing hand, together we extinguished the fire. I was later complimented on my action, and I am happy to say that my kindly mentor still survives and occupies an honoured position in the department.

Out of the memories of my many years' experience of fire-fighting it is difficult to select one particular conflagra-tion as being more thrilling in its incidents than any other. All fires entail risk to life in a greater or smaller degree, and are therefore replete with that human interest which makes special appeal to the heart. For even in the factory or warehouse outbreak human lives are endangered; the lives of the firemen employed. But sometimes cir-cumstances do arise which require the pen of a Stevenson to give them that actuality and force which alone can depict them in their fearful vividness.

THE GENNEVILLIERS GAS-WORKS FIRE BRIGADE
LIFE-SAVING TRUCK.

To my dying day I shall never forget the horrors accompanying the burning of the Park Avenue Hotel.

At 1.30 on the morning of February 22nd, 1902, the gong in the quarters of Engine Co. 72 sounded 3-3-446, which, translated into bald English, signified the fact that a dangerous and threatening fire was raging in the vicinity of Park Avenue and Thirty-fourth Street in the Borough of Manhattan, New York City. In other words, it was a third alarm, summoning to the scene thirteen engine companics; four hook and ladder companies, the Chief of the Department, the Deputy Chief, and four Battalion Chiefs. Engine Company 72 responded on the third alarm and in less than twenty seconds after the receipt of the first tap of the gong they were clear of the doors of the quarters and on their way to the fire. At that time I was Captain of this company, and beginning to feel the full weight of my responsibility. A fierce gale from the northeast raged about us as we left our comfortable quarters, the snow and sleet lashing our faces and making vision almost impossible. The driver of the engine has since often assured me that for a mile and a half of the distance to be covered he let his horses gallop without knowing his precise whereabouts. Yet in spite of the storm we reached the scene of the fire in less than five minutes.

On our arrival we found that the 71st Regimental Armory, situated at the southeast corner of Park Avenue and Thirty-fourth Street, was ablaze. The interior of this imitation fortress was of wood and filled with arms and ammunition of every description. Evidently the fire had been burning for some time, for as we pulled up there was a constant rattle of exploding cartridges, for all the world as though our services had been requisitioned to a field of battle. In addition to this the building was heavily charged with a smoke, which reached the explosive point as soon as an opening admitting a fresh supply of oxygen was effected. Orders were received from the commanding officer, Deputy Chief Duane, that a line was to be taken into the armory by the Thirty-

fourth street entrance. At this moment the truck companies succeeded in opening these doors, but the pressure
of heated air and gas blew the men back into the street.
Almost instantly the whole interior of the building was a
seething mass of flame. Nothing further, apparently, could
be done here, as my instructions then were to cover the
dwelling houses on the east side of Thirty-fourth Street,
where we fought the fire back until the wall of this part of
the armory fell outwards, burying our line and cutting it in
two. Some idea of the difficulties confronting us can be
imagined when I add that the position from which we
were fighting consisted of a narrow strip of street, some
twenty-five feet wide, bounded on one side by the flames
and on the other by a trench forty feet deep, which was
being prepared for the reception of the present subway.
The break in our line naturally shut off the stream, and
I went immediately to see what had happened. Meeting
the officer in charge, I was ordered to take yet another
position in Park Avenue, in order to cover the Fourth
Avenue car stables. These were to the south of the fire
and it was this change which brought my company into
a position that enabled it to assist in the most harrowing
and exciting events that I have ever experienced on land
or sea.

To begin with, this manœuvre necessitated our crossing
the subway trench, which, incidentally, we were told, contained three tons of blasting powder. It has always been
a marvel to me that this did not explode, exposed as it
was to sparks and burning embers. We managed to reach
our goal in safety by means of the engineering shores used
in the "cut and cover" system of excavation.

At this moment, from some unexplained cause
the Park Avenue Hotel took fire. The figure of
a woman clad only in her night clothes appeared
at a fifth-story window and above the roar of the
flames and the exploding of the ammunition could
be heard screaming for help. Even as her voice rang
out, guests could be seen watching the conflagration from

their bedrooms, while in the foyer men were strolling about, cigars in their mouths, discussing with interest the probable amount of damage which would be caused by the blaze. Little did they realize that the angel of death, with wings outstretched, was hovering over the building in which they were. Our change of position made us among the foremost to effect an entrance. From the first we were hampered by the revolving doors, which prevented our handling our lines with facility. Thus valuable time was lost and our task rendered the more difficult. Our arrival had been heralded with the frankest incredulity, but once onlookers realized the grizzly danger threatening their dear ones, they had to be forcibly restrained from adding themselves to the human sacrifices awaiting us upon the floors above.

As we climbed the stairs the smoke grew denser and denser, till our breath came in strangling gasps and physical endurance seemed about to fail. It was impossible to see. On hands and knees we groped and felt like blind men, instinct our only guide.

And then the horror!

Imagine crawling sightless along a strange corridor. Imagine the outstretched hand wandering over an unknown substance which slowly reveals itself to be a corpse. That would be a ghastly situation. But add to it the distant crackling of flame licking its way remorselessly from floor to floor, the shouts of firemen in difficulty, the sobs and piteous entreaties of unseen women dying slowly from suffocation; and can Hell be pictured as more hideous!

Grimly, however, all ranks alike stuck to their lines, scrambled over these gruesome barriers and with almost miraculous tenacity of purpose succeeded in quelling the grim destroyer. As a matter of fact the whole outbreak was under control within a short time and it was then possible to realize the tragic uncertainty of life. For had the men and women, whose lifeless forms encumbered the passages, only remained in their rooms, not one need have been lost.

As we returned from the holocaust and passed through the front hall, it seemed incredible that even then there were those who were still sceptical that Death, the Reaper, had passed with his scythe. But next day, the unfortunates in the Tombs prison knew of the harvest, for amongst those who had fallen in the mowing was one whom they called their angel, Mrs. Foster, the Florence Nightingale of prisoners.

No lives were lost in the Armory fire, but the number of persons who perished in the hotel amounted to twenty.

It is naturally impossible to lay down hard and fast rules for the guidance of people who are unfortunate enough to be caught in such fires, but broadly, the safest course to pursue is to avoid the vicinity of elevator shafts.

Perhaps I may include amongst these few stories an incident so commonplace to the fire-fighter that it was never even officially reported, but which should bring home to the outsider the daily unconsidered risks accepted by the former without demur.

On this particular occasion the Captain of our company received orders to take his line to the roof of the building to the north of the one on fire. The intention was to breach the wall of the burning structure with battering rams, in order to better attack the flames. As our point of vantage was some fifteen feet lower than the top of the wall to be attacked, this move was excellent strategy. We lowered our roof rope to the street, where it was made fast to the hose and hoisted up to be in readiness. In order to make it perfectly secure I was instructed to lash it at the cornice of the roof with a special knot, known to firemen as a "rolling hitch," preparatory to starting the water. Properly to adjust the knot it was necessary for me to lie at full length near the edge. I had just got a turn of the rope round the hose, when a warning cry caused me to look around and I saw all hands running for the north coping. There was no need to tell me that the wall was falling out and I jumped to my feet, letting go of rope and hose. By great good luck I escaped becoming entangled with them or I should have been dragged to my death. Just as I reached my comrades, the wall crashed down, carrying with it the roof of our building and the fire instantly swept into the rooms beneath us. It then became imperative that we should reach the next house by hook or by crook or perish. Between us and safety was a "pocket," that is to say, there was first a drop of some fifteen feet on to tiles, followed by a climb of the same height up a bare wall. This latter appeared to offer an insurmountable obstacle, but the fire was hard on our heels and desperate men reck little of seeming impossibility. One of our number, a giant in stature and strength, named Michael Byrne, raised me on his shoulders and like an acrobat I placed my feet in his hands, making our combined length almost the height of the wall. With a slight spring I succeeded in clutching the top of the coping and with sailor-like agility I hauled myself up. Finding a short ladder on the roof I passed it down, by which means the others escaped, though the captain, the last to leave his post of danger, was badly burnt about the face and hands.

While there must always be difficulty attendant upon the fighting of fires, as can be imagined, those that occur in the winter months are by far the most physically trying. For instance, during the great blizzard of 1888, which paralyzed all traffic in New York, my company was summoned to a fire. All telegraph wires were down, and the alarm was brought in by a mounted messenger. On leaving the quarters we found the streets nearly impassable, and after an odd hundred yards our apparatus became stalled. We then commandeered any horses we could find, and pushing and pulling we worked our way through the snowdrifts to within three hundred yards of the outbreak, where the engine pole snapped in two. We left the latter where it was, but succeeded in securing sufficient hose to be serviceable, and for thirty-six hours we remained on duty without food or rest.

Again it sometimes happens that the fire hydrants become frozen and precious time is lost in thawing them, though now-a-days this occurrence is becoming increasingly rare, owing to the improvements introduced in modern water supply. But in northern latitudes King Frost is the *bête noire* of the fire-fighter, and must be held indirectly responsible for some of those catastrophes which occur during his reign.

In concluding these brief personal reminiscences the writer hopes that he has shown in a straightforward way what the life of the fire-fighter really is, the stirring incidents which compose it and the great possibilities therein for young men of enterprise and ambition. Some months ago the whole civilized world was stirred to its depths by the tragic and glorious death of the British explorer, Captain Scott. It has been said and said rightly, that the world is the better for the man and his example, which will live through the ages and doubtless will serve to stimulate others, when called upon to face great crises. And the writer ventures to say with all humility that the fireman hero, though unknown to history and unsung in legend, meets death as bravely and dies as gloriously in the service of his country and his people.

CHAPTER V

THE history of the Paris fire brigade is of exceptional interest and well deserves study. Its early organization and manifold developments were contemporary with the principal change of thought and government in France, and to a certain degree echo the tendency of different forms of State control favoured in that country during the past two centuries.

In the year 1716, the city of Paris organized its first regularly constituted fire-fighting force. This consisted, so it is stated, of thirty-six manual engines with a personnel of forty to operate them. By 1785 the personnel had increased to three hundred, and in 1789 the first fire regulations were issued. The year 1807 saw the force placed under the command of the Prefect of Police and the introduction of the brass helmet, which is still worn by brigades of distinctly conservative tendencies, notably the London force.

Thus, whilst originally a civil organization, in 1811 it was turned into a military corps, and in 1867 it was advanced to the status of a regular regiment commanded by a Colonel and consisting of two battalions of five companies each. In this formation it remains today, with the slight difference that now each battalion numbers six companies, its official designation being *Le Regiment des Sa peurs Pompiers.*

By its constitution, the Paris fire brigade is something more than a purely fire-fighting force. In times of disturbance or war it may be called upon for military duty, though it is difficult to see how the fire risks of Paris could be guarded were such a step ever taken. However, though under the military authorities as far as recruiting, internal administration, discipline, promotion and punishment are concerned, for fire purposes it is placed under the direct orders of the Prefect of Police, whose wishes regarding all technical matters such as scientific training, fire mobilization and equipment are paramount. Add to this that the City of Paris is financially responsible for the entire expense of the regiment and it will be seen that there are no less than three interested parties in the maintenance of the corps.

Hence, in order to avoid confusion and friction, there is a joint committee formed of members from these administrations, which settles all questions involving its common interests.

The present strength of the force consists of the Colonel Commanding, forty-eight officers, four medical officers, and 1,803 non-commissioned officers and men. Of the latter two hundred are sergeants, three hundred and sixteen are corporals, the balance of 1,287 being rated as firemen.

As a rule officers are recruited from ordinary infantry regiments, entering as sub-lieutenants, but they are first obliged to pass a medical and technical examination before a special commission. If successful, they then undergo a course of fire service instruction, and are required to attend all important fires as spectators in order to familiarize themselves with the actual handling of apparatus. No doubt it is easy to be hypercritical, but to the scientific fire-fighter, this appears to introduce an element of weakness. The marine engineer officer does not learn his calling by watching the efforts of others any more than the surgeon is qualified to operate upon a patient because he has had the chance of observing the greatest masters of the knife. It cannot be too strongly emphasized that fire-fighting is a science which demands of its students that they should understand its complexities from A to Z, and this can never be accomplished by any amount of theoretical schooling.

To this extent, then, it may be questioned whether the training of the officers serving in the Paris fire department is of the best for practical purposes.

Non-commissioned officers, who, under the conscription law, may elect to do their service with this corps, are not required to pass the technical test should they wish to remain with the regiment. Senior non-commissioned officers rank as warrant officers and as a rule serve for twenty-five years, while corporals and firemen are limited to fifteen years' service, then retiring with a pension.

The regiment is recruited principally from artisans, builders, laborers, mechanics and coachmen, the idea presumably being that most of the running repairs and a certain proportion of constructional work should be carried out by these men in the workshops of the brigade. Now this system is also open to comment. A firemen should be first and foremost a fireman. The last thing that should be made of him is a Jack of all trades. His calling is of the most strenuous and when not actively engaged at fires he has plenty to do in seeing that his apparatus is in proper condition. To set him to construct the body of a departmental automobile or to repair a major defect in a pumping engine, is to remove him from his proper sphere of operations, and since science has not yet solved the problem of keeping one man in two places at the same time, the actual fighting units must be proportionately weakened.

The pay of the Paris firemen, according to American ideas is so small as to seem ludicrous, but it should be remembered that it is based upon the army scale, which in all European countries is framed upon as low a basis as possible. It commences roughly at thirty cents per diem, rising to forty-four cents should the fireman gain the rank of corporal during his three years under the conscription law.

Otherwise the pay of those proposing to qualify for a pension ranges from $275 to $325 per annum. Free quarters are provided for the married and unmarried non-commissioned officers and men, as well as lights, fuel and uniforms, but no messing is included. Regular firemen get 30 days leave annually, but conscriptionaires only 15, there being short leave once a week for all ranks.

It will be seen from an examination of the following table that the area of Paris has increased in the proportion of 1 to 2.26 since 1841, and that the number of fires is ten times as many as at that date. During the same period the strength of the brigade has only been doubled.

Year	Brigade strength	Area of Paris	Population	1 fireman for each—inhabts.	Expenditure	Number of fires	One fire every —hours
1841.......	803	13.26 sq. mls.	935,260	1,145	$ 146,745 £ 29,349	203	43
1857.......	889	13.26 sq. mls.	1,278,705	1,438	$ 169,380 £ 33,876	298	29
1860.......	1,208	30.2 sq. mls.	1,537,486	1,241	$ 208,495 £ 41,699	445	19
1867.......	1,498	30.2 sq. mls.	1,848,075	1,233	$ 295,525 £ 59,105	690	12
1879.......	1,690	30.2 sq. mls.	2,126,230	1,258	$ 364,620 £ 72,926	878	10
1910.......	1,803	30.2 sq. mls.	2,763,393	1,532	$ 721,400 £ 144,280	2,030	4.20 min.

The statistics in the last annual report show as stated above that the brigade attended 2,030 fires, exclusive of chimney fires, which numbered 1,554. They also rendered various additional services amounting in the aggregate to four hundred and forty-four calls. These last were exceptionally numerous during that year owing to there being many cases in which assistance was given in connection with floods in Paris, work in which the corps has always especially distinguished itself. Over and above these legitimate calls, the department responded to no less than seven hundred and twenty-seven false alarms.

Before going into a detailed description of the equipment and work of the brigade, it may not be amiss to point out certain factors in connection with its constitution which will enable the lay reader the better to appreciate the vital part this force plays in fire protection throughout the whole of France. Owing to the number of young men who elect to do their military service in its ranks and who, at the end of their allotted time pass out into civil life, there are all over France many serving in rural and provincial forces who have thus acquired a considerable amount of useful experience and whose influence must be advantageous in the development of local fire control. In fact, it is worthy of notice that the Paris fire force is regarded by the authorities as something more than a municipal institution; rather is it intended to meet national requirements. Thus, as a rule, a squad of sailors from the French Navy is attached for an instructional course of six weeks, as it is felt that opportunity should be given to all in Government employ to acquaint themselves with the rudiments of this science.

That the constitution of the Paris Fire Department is clumsy cannot be denied, since it is under military, police and municipal control; yet the introduction of military influence may perhaps be regarded as beneficial. A certain prestige attaches to any form of military control, and in this instance has caused this force to be looked upon as something in the nature of a *corps d'élite*. Broadly speaking, whilst a brigade, which is essentially a municipal institution, may develop a tendency towards loss of status and lack of discipline in its truest form, owing to political, party or labour influence, yet it seems the most logical form of organization. But at the same time, fire control is a question of such serious moment that some form of governmental ascendancy in the hands of a competent central authority appears to be beneficial, if not absolutely necessary. At any rate, the basic structure of a modern fire department should be moulded along semi-military lines, for even as on a battlefield success or failure, victory or defeat, may be largely determined by the unquestioning obedience of all ranks to their superior officers, so when fighting as crafty an enemy as fire, it requires not only the skill of the commander but also confidence and prompt compliance with orders on the part of subordinates. This can only be engendered by a quasi-military training, such as it has ever been the ambition of New York Fire Chiefs to inculcate into the force under their command. There are, no doubt, many excellent fire brigades controlled wholly by municipalities, but there are also many bad and inefficient ones, which apparently satisfy ignorant and incompetent local authorities.

Fire prevention in Paris itself is practically looked upon as an administrative precautionary measure initiated and applied by the Prefect of Police after consultation with the officers of the Paris fire brigade, with the municipal technical officers and with such other parties as have some concern in the matter.

Amongst the general public, the architectural, engineering or surveying professions, and even in governmental circles, little or no interest is manifested in the question. There is, in fact, no body either in Paris or France framed along the lines of the British Fire Prevention Committee, which is representative of technical opinion and is formed with the express intention of formulating precautionary rules. Of course, after some great disaster an irresponsible clamour for precautionary measures must needs arise, and in this particular Paris is no whit different from New York or London. Fanned by a sensation-loving daily press, blame is scattered broadcast, quite irrespective of equity, and the simple necessities of the situation are swamped by the volume of hysterical and irrational vapourings poured forth by the ignorant. And, be it added, like most press sensations, the matter is speedily forgotten and nothing permanent eventuates. Such agitation arose after the disaster at the Paris Charity Bazaar, after the burning of the Iroquois Theatre in Chicago, after the destruction of the Exeter Theatre in England, but curiously enough until very recently it never resulted in any organized effort on the part of members of the public possessing technical knowledge to combine and assist the authorities on the subject.

Thus for practical purposes it will be seen that the Paris

Prefect of Police embodies all the initiative, which should be provided by a French Bureau of Fire Control. At the same time, however, he is possessed of two great advantages, which enable him to use his position of amiable autocracy to the fullest extent, namely the funds and the personnel wherewith to investigate matters, to undertake tests, and to enforce by means of administrative order such safeguards as he may see fit to demand. But equally, the lack of public interest in fire prevention places him in a most unenviable predicament, inasmuch as having no private scientific society or public commission to take the initiative and demand certain safeguards, he is necessarily compelled to act on his own discretion.

Hence, were it not that Mons. Lepine, the world renowned ex-Chief of the Paris Police, was a man combining the greatest strength of character with an iron tenacity of will, it is no exaggeration to say that that city would be one of the worst equipped of the great modern capitals as far as fire control is concerned. In this connection must also be noted the names of Mons. Lepine's most able and competent colleagues, Colonel Vuilquon and Lt.-Colonel Cordier.

For effective measures of fire protection, Paris is divided into twenty-four zones, the size of which is governed by the density of the population. In each of these zones is a fire station with which the fire alarms are connected. Each station is equipped with not less than three fire-fighting appliances; a motor-propelled steam pump, long ladder and hose wagon. In addition, the four stations situated in the most populous sections are provided with an electromobile first aid machine, fitted with a small electrically-driven pump. These machines are intended to deal with a fire in its first stages, much in the same way as the chemical engines, common to American fire practice. The number of men on duty at each station consists of three non-commissioned officers and twenty-six corporals and firemen.

In the event of a call the fire station notified immediately sends out one or two appliances, and at the same time

telephones the next nearest station to the scene of the outbreak. While stations help each other in sending on appliances to calls, they do not as a rule deplete their apparatus to a dangerous degree. The pump and ladder are always employed for the defense of the zone belonging to
a particular station and are always the first upon the scene.
Fifteen men form the crew of the motor engines, which also
carry three hose reels with 1,968 feet of large hose, 525
feet of small hose, smoke helmet and air bottles, life saving
lines and a ventilator. These machines are of forty-five to
sixty horse-power, and have a centrifugal pump, which
can deliver 660 gallons a minute. Horse drawn steamers
carry 1,575 feet of hose, short ladders and all the necessary gear for coupling up with hydrants. Whether shipped
on motor or horse vehicles, the long ladder can be extended
to a height of sixty-five and one-half feet. Motor traction
is being rapidly introduced, and at the present moment
there are in the brigade forty-nine automobiles, of which
eight are electrically propelled. In addition, seventy-six
horses are hired by contract on the understanding that they
are entirely at the disposal of the department. The contractor furnishes fodder and bedding for the horses as
well as the necessary harness and stable gear, for which
he receives eighty-three cents per horse *per diem*. The
training of these animals is good, though being of Flemish
breed they are too heavy for the dashing work accomplished by many other fire brigades.

Paris possesses seven reservoirs which supply its fire hydrants, the installation of which was commenced in 1872.
These latter now number 7,726, and when the system is
complete they will be about a hundred yards apart. Their
nozzle pressure varies from fifteen to seventy-five pounds,
according to the height of the reservoirs, and usually the
average pressure is sufficient for working purposes. Besides
these, there are six hundred and ninety-one hydrants belonging to public buildings or private firms. The brigade
can be called firstly, by five hundred and twenty-one alarm
boxes situated in the public streets; secondly, by four hun-

AMBROTLAS

dred and nmety-five private fire alarms (theatres, public buildings and so forth), and thirdly, by the use of the police or public telephones. All such alarms are of practically the same pattern. They consist of a square box upon a pedestal, instructions for operating being printed upon the glass front of the apparatus. On breaking this glass, the door automatically flies open, making a contact which rings a bell in the fire station. Contained within the box is a telephone transmitter for the purpose of giving the station the address of the fire. When the message is understood a buzzer is sounded to signify that fact. The disadvantage of this system is obvious, inasmuch as there is no check upon false alarms, while in moments of great emergency the individual is only too inclined to bungle anything in the nature of a telephone message. In the event of no message being received the appliances proceed to the neighborhood of the fire-box.

Great attention is paid to the physique of the men forming the corps and in addition to the squad and company drill, which constitutes part of their military training, considerable time is devoted to gymnastics. It may be open to question whether the practice of what, for want of a better term, may be called acrobatic exercises, really improves the stamina in the most advantageous manner, yet a feat such as the following possibly inspires a certain amount of self-confidence.

The apparatus employed is called the piano. It consists of a vertical timber structure, about fourteen feet high, comprising a number of horizontal boards separated by a groove, to imitate the rustic grooving in classical architecture. The men ascend this with their fingers alone, "jumping" each board with their two hands simultaneously. These grooves, which form their only support, are but one and a half inches deep. Incidentally, last year the men of the brigade won the regimental cup for gymnastics, open to the whole French army.

Regarding a fuller account of the apparatus in use in the Paris fire department, it need only be said that with the exception of some unimportant particulars, the appliances are much the same as those employed in the New York fire department, a detailed description of which will be found in the chapter under that heading. It only remains to be emphasized in instituting comparisons that the utility of apparatus depends solely upon its suitability to its environment, and the narrow streets of Paris offer an insuperable obstacle to the giant appliances in use elsewhere.

The Paris fire brigade having found that considerable delay was caused by the summoning of a building contractor when dangerous walls, etc., required attention, and that the risk thus incurred by the men during the delay preceding his arrival was considerable, decided to provide itself with its own gear for dealing with dangerous structures. This consists of eight "horse traps," two comprising an unit and manned by fifteen men. Those employed in this particular squad are all carpenters or builders by profession and are thus supposed to be in a position to render first aid to any building in peril of collapse, with facility and expedition. Frankly, this feature in the brigade is one also of doubtful value. True, the numbers allocated for this particular service are not excessive, but in dealing with problems of a similar nature in New York the author has found that in such cases of emergency it was more satisfactory to count upon the services of building contractors of known standing than to rely upon a small subdivision of the fire corps itself, which, from the nature of the case, cannot possibly possess the scientific and architectural skill necessary to cope with such a vast and intricate question as the shoring up of a wall in momentary danger of collapse. Wrecking crews are employed in the New York fire department, but their duties are very much narrower than those of their French colleagues.

An account of a fire in Paris, drawn from the report of the British Fire Prevention Committee's Journal, may not be without interest to readers, lay and professional alike.

The site of the outbreak was a linoleum factory situated in the Rue de Vouille, a long, narrow street approached by

thoroughfares at either end, and backed on one side by tenement buildings and on the other by a railway. Obviously
the chief risk was that the fire might spread to the tenements, and hence the main attack had to be made from
either end. Seven motor pumps were brought into operation, supplying 13 jets, while two more were worked from
a hydrant. The number of officers and men employed numbered 135. It only remains to be said that the disposition
of the apparatus as evidenced by the plan reproduced below was admirable, the officers in charge having clearly and
.quickly grasped the danger zone, and it is satisfactory to
note that the blaze was under control within 90 minutes of
the arrival of the first engine upon the scene.

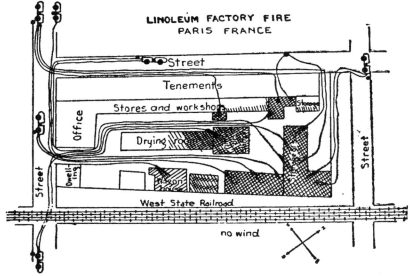

Owing to the part played by the Prefect of Police in the
control of the Paris fire brigade it is natural that some
form of coöperation should exist between the two departments. In fact it is not too much to say that herein lies
a connection of considerable value. For ordinary small
fires the mobilizing of the police necessary to keep the
ground is dealt with by the provisional police superintendent in whose area the fire occurs and he can also draw
assistance from neighboring divisions. In the case of large

fires, police headquarters sends immediate aid from its reserves, including if necessary republican guards and strong cyclist sections. The principle of having a number of police ready for immediate turnout on bicycles to any point in the city is both expeditious and advantageous and merits more than passing attention from the authorities of every large municipality. Upon coöperation between the police and fire departments much depends, and it is only by constantly playing into each other's hands that the former can rightly judge how far away a crowd must be kept for their own safety's sake, and in order that the efforts of the firemen may be unhindered by the ill-judged incursions of the curious.

For fire protection along the front of the river Seine there is a special organization known as the *Brigade Fluviale* or River Police. This consists of a chief inspector, four assistants, and thirty-six policemen, twelve of whom are pilots and mechanics. Not only does this force serve as auxiliary to the fire department, but it is trained for emergency work in times of flood as well as acting as police in the usually accepted sense of the word. Needless to say, owing to the tortuous narrowness of the Seine, the apparatus in use is small, but it is serviceable enough for its purpose, and the men in the corps are in addition expert life savers of drowning persons.

It is a curious anomaly of this command that its hours of duty are only from 7 a. m. till 10 p. m., and hence it is practically unavailable for any night emergency. The authorities, who are responsible for this incongruous state of affairs, must evidently possess a touching confidence in the designs of *le bon Dieu,* as, to the ordinary ideas of the fire-fighter, the hours of most danger are precisely those when, it is to be supposed, the officers and men of the *Brigade Fluviale* are wrapped in slumber.

Salvage work in Paris is carried on by a distinct section of the fire department, and is in no way reliant upon any outside or independent assistance. This whole question **of the** interdependence of the fire department upon a

private salvage corps, and vice versa, receives careful consideration in another section of this volume, but none the less it may be broadly stated that there are advantages attaching to an undivided control of both these departments, and since Paris was the first to adopt such a measure a short account of its equipment for that purpose may not be without interest.

The Paris salvage service commenced operations in 1904 and is intended to limit, as far as possible, the damage caused by water or fire to all kinds of property. Since salvage duties form part of the ordinary duties of the brigade, instruction in this special branch is given to all ranks. Special appliances for the purpose are placed in six stations, each of which has a certain number of zones to protect, but the appliances of one area may be sent to another according to the severity of the fire. Besides this, every one of the twenty-four motor apparatuses of the department carries some salvage gear, so that a proportion of this work can be accomplished without the presence of the special cars.

Each salvage car is manned by one non-commissioned officer, two corporals, four firemen and a driver. On arrival at a fire the man in charge of the appliance takes his orders from the senior officer of the fire-fighting force, who employs his services as he thinks best. Thus, in case of emergency, the men of the salvage corps can assist in the fire work or the men employed in the fire work can assist in the salvage. The salvage units comprise six motor cars with a wide radius of action, each carrying a crew of eight men, the whole being under the charge of a superior officer and each carrying no less than fifty ordinary covers, fifteen special covers, one special scaling ladder, one step ladder, one set of draining gear, and a large supply of mops, brooms, swabs, sponges, trays, small covers, ropes, lamps, axes, carpenter's tools, bags of sawdust, telephone fittings and so on. There is also a reserve car, and in addition to this every motor pump carries two covers and other minor salvage gear.

Towards this comprehensive service the insurance companies pay $40,000 per annum, and in addition nominate two of their officials to do service if called upon in a technical or consultative capacity.

Probably the Paris fire department is the only one in the world which can bring so effective a plant to the seat of operations so .quickly and with so little delay, and broadly speaking it is without doubt an advantage to have at hand so large and competent a force upon which to draw at a moment's notice. The personnel of both fire brigades and salvage corps is, after all, only human and it is impossible always to avoid some friction between the two bodies, when each has a different object in view and is naturally anxious to look after the best interests of their respective paymasters. It is in this direction that Paris benefits.

Generally speaking the Parisian theatres can scarcely be said to make any special claim for excellence in either architectural construction or equipment. But in this respect they differ little from those of other countries, which except in rare cases seldom come up to the standard of modern requirements. In the case of Paris, the majority of the buildings are old and the proprietors have vested interests, necessarily rendering any action on the part of the public authorities a difficult and thankless undertaking. Nevertheless, a systematic effort may be observed on all sides to ameliorate the dangerous features in these old buildings and to ensure safety of the audiences·as far as is practicable under existing conditions. The primary features of the protective system observed in Paris appear to be very similar to those in vogue in New York and consist in the installation of a fire-resisting curtain, large ventilator openings, absence of rubbish, the non-inflammable treatment of scenery, constant inspection and lastly the organization of "fire-watches," composed of regular firemen, who shortly before every performance make a round and test the˙fire appliances, remaining until the conclusion of the entertainment, when the appliances are once more put under trial. Also, as in New York, plans for new theatres are inspected and reported upon by the fire department.

All of this is most satisfactory and is evidence that the French authorities are keenly aware of the terrible fire risks in theatres where even a false alarm may result in a hideous and unnecessary loss of life. But with the National Opera House of Paris another tale has to be told, and it is literally amazing that its equipment and construction should be such as to make even the most uninstructed in the peril of fire pause and hesitate.

Granted that the foundations of this historic pile were laid as far back as 1863, yet owing to the Franco-Prussian war the building was not really completed till some twelve years later, the opening taking place in 1875. There has always been an idea that buildings of a period antecedent to our own day of rush and hurry were more substantial and of better construction than the jerry built shacks of the modern real estate agent, who hides the worthlessness of his wares under liberal coatings of gilt and gingerbread. Yet, judging from the Paris Opera House, the architects concerned in its erection must have counted fire as one of the negligible happenings of fate. No less a sum than $7,500,000 was lavished upon the building, but apparently the imagination of the gentlemen responsible for its erection only carried them as far as architectural magnificence, and they were blind as to such matters of minor importance as the safety of audience and artists. But again perhaps it is too much to expect that an architect of the sixties in the last century should have realized that "panic bolts" to doors, rounded corners, and continuous handrails formed safeguards for human life. One might legitimately expect, however, that such precautions would have presented themselves to the minds of the present day directors.

To quote from the report of the British Fire Prevention Committee issued after their visit to Paris, "The Opera House stage is generally considered to be one of the most dangerous, if not the most dangerous, in Europe. It is

mainly of wood construction, supported in parts by unprotected cast iron columns. It is a mass of old-fashioned
windlasses, pulley gear and a veritable forest of rope.
Little can be said, beyond that it should be entirely gutted
as was the case with the Royal Opera House, Covent Garden, and that a modern stage should be fitted in its place.
The safeguards, however carefully devised, are discounted
by the highly inflammable and complex character of the
stage equipment."

It must be clearly understood that this excerpt is given
with no idea of disparaging one of the great art centres
of the modern world, but only with the object of bringing
home to the ordinary citizen the fact that with all the history of fire disaster behind them for their guidance those
responsible for the safety of the public, unintentionally, no
doubt, even today regard the subject apparently as not one
of serious import!!! Further comment is surely needless.
Suffice it to say that judging from official reports concerning the fire equipment of the Paris Opera House, it is ludicrous, were it not possessed of its tragic side. Its structural height is nominally 13 stories, the fire protection of
which is served by three mains, the high pressure being naturally for the protection of the upper part of the building.
Considering that the maximum pressure off the mains is
only 70 pounds, it is difficult to see how even the tenth story
could be protected, let alone the thirteenth.

The subject of fire protection in department stores on the
other hand, has for some time past been receiving the careful consideration of the Paris Fire Bureau, and in this connection the "modus operandi" of the great Bon Marché
stores offers an example worthy of imitation by many similar establishments in big American towns. This firm
maintains a private fire brigade of 41 men, who do nothing
else except watch and fire duty. One-third of this number
sleep on the premises, while to assist them is a special staff
of 18 night watchmen. A portion of the regular sales staff
also is instructed in fire duties, being especially trained to
deal with customers and others in the case of a fire panic.

The store possesses its own water supply and sprinklers are fitted in all parts of the building considered to be particularly dangerous. Great care is also taken over the collection of waste paper and rubbish generally. It is gathered into sacks and removed to a fire resisting room in the basement, which is lighted from without, is supplied with sprinklers and possesses a self-closing iron door. In addition, all packing material is stored in a special apartment, and is only issued as required. The elevator shafts are taken above the roof, the upper part of the shaft being glazed with thin glass, the idea being that in the event of fire the heat and smoke should go well clear of the building. Finally smoke helmets are kept ready for the slightest emergency, each being fitted with a portable electric bulb and a supply of oxygen sufficient to last 90 minutes.

At different periods during the last century, notably in 1851 and 1858, efforts were made by the Government of the time to obtain some form of provincial fire service on national lines, whereby the responsibility of the different local authorities might be centralized. These efforts met with scant success, although a number of Communes formed fire brigades as sections of the national civic guard. The first modern decree on the organization of French fire brigades was signed by Marshal MacMahon in 1875, and comprised 35 articles setting out the requirements and conditions of service in great detail. Of course it was in itself only applicable to the day of the manual engine, but even now it can well rank as a model to all countries as a code which nationalizes a necessary service, which is all too easily allowed to remain unrecognized where the independence of local authorities has become a veritable fetish regardless of the best interests of the community.

The following are some of the features of the decree of November, 1903, which today governs the formation of Communal fire brigades in France and marks a stage in the development of the old decree of '75. Fire Brigades are primarily formed to do fire service, but may also be called upon to assist in case of any serious accident or catastrophe.

If they so desire, and with the permission of the Home Secretary, they may be armed, but under those circumstances they are not allowed to carry their rifles outside the limits of their own district. Fire brigades can only be formed with the sanction of the President of the Department, after proof has been given that sufficient appliances exist for the brigade to man, and that means are available for the purchase of uniforms and the general upkeep of the force in an efficient condition for a period of at least 15 years.

The general organization is along strictly military lines, men being enlisted of their own free will for a period of not less than five years. Officers are appointed by the President of the Republic on the advice of the Prefect of the District or the Mayor of the Commune. Their rank is ex-officio military.

But the chief point in this connection is the effort which has been made by the various provincial fire departments towards a common federation of all brigades, the standardization, so to say, of the system as a whole. The objects of this federation are to improve the French fire service generally, conduct assemblies, competitions and exhibitions with a view to encouraging the ambition of various local units and to the creation of a species of local *esprit de corps*. Incidentally also, comprised in the scheme is a plan for benefiting those who are injured in the course of duty and of assisting their wives, widows, or families. At the present time this federation consists of over 104,000 members and there is no reason why, if managed along normal lines and those of least resistance, i. e., in conjunction with the governmental authorities, this Federation might not prove of inestimable benefit to all concerned.

The competitions conducted are of peculiar value, as they do not consist of events which are merely a matter of athletic celerity, but are rather founded upon a semi-scientific basis. By this means successful brigades may be regarded as not only occupying the position of merit allotted to them in any particular competition, but as embodying thereby their actual standing in the ranks of the provincial fire department in France as a whole. There are also theoretical examinations for officers, which are taken separately and of a graduated character, there being five groups, A to E, admission to the higher group having to be preceded by the obtaining of honours in the next lower group.

In fact, it is not too much to say that this Federation, although in need of modernization as regards some of its details, is generally beneficial in the highest degree to the French provincial fire service, and by engendering enthusiasm and a spirit of emulation has done much to advance the cause of fire-fighting in that country. Monsieur Guisnet, its President, has admittedly a difficult body to control. Political influences and administrative problems of importance have to be constantly overcome and adjusted with that diplomacy which alone can bring success to any organization of such magnitude. And, though, from time to time setbacks occur and attempts are made to discredit the work accomplished, the fact remains that the very genesis of such a union is a hopeful presage for the future. Were it possible to train the members of all fire departments in a country along national lines in a similar manner to that in which the apparatus of various cities in the United States has been standardized, then without a doubt a great step would have been taken forward in the science of Fire Control.

Admittedly, of course, in a vast country like the United States, such a scheme is impossible of realization, but in smaller areas, such as England and other European countries, the idea would certainly appear to merit consideration.

CHAPTER VI

As might be expected by those conversant with Teuton thoroughness, the question of fire control in Germany has received the most careful consideration on the part of the authorities from the Emperor and Empress downwards. This has resulted in the centralization of all executive authority, which in itself possesses many advantages. In Berlin, all matters relating to building construction, factory inspection, the storage of inflammable material and other details of a similar nature are under the supervision of the Berlin Royal Police, with which the Fire Brigade is incorporated. The advantages of this system are obvious. Thus a factory inspector, a superior officer of the fire department, a superior officer of the sanitary police and the police building surveyor frequently work together, and confusion as to responsibility or the overlapping of various forms of control is eliminated.

Now, admittedly, this system is excellent, but since prevention is better than cure, great efforts are made to instil into the minds of children at an early age the necessity of exercising great care in the use of matches, lamps, candles and open lights. Towards this end special courses are arranged in the public schools, whereby boys and girls are taught by fable, picture or simple instruction the dangers inseparable from imprudence in the use of the above mentioned articles. These simple educational methods are having a most marked effect on the whole of the coming generation in Germany and fatalities from burns amongst young people have decreased, while their parents also have

64

grown more cautious. Naturally the full results of this teaching will not be felt for another ten years, when its effect upon the incidence of fires should become marked.

Building construction in Germany generally is of a solid and substantial nature, both as regards business and residential premises. The interiors, being subject to the inspection of the local building control department, risks such as those commonly met with in tenement houses are avoided.

The centralization of fire control has also had important results as regards the high standard of safety existing in most German theatres. This supervision is responsible for the introduction of the specially heavy fire curtain in general use, and for the installation of a system of stage lighting, which does away with the more dangerous features of the older methods. In this connection it may be noted that the theatre owners find the police restrictions in no way irksome, even though that most unpopular official the "Censor" is also a member of the department.

One final feature of the Prussian brigades merits attention. The duty for firemen is so arranged that after forty-eight hours at the fire-station, they are entitled to twenty-four hours rest at home. During their period on watch starting at 8 in the morning, they are actively employed till 10:30 P. M., when, unless summoned to a fire, they may sleep till 6 A. M. On that day they are relieved from 2 P. M. till 4 P. M. in order that they may take part in the night watch without undue fatigue. There are no married or single permanent quarters for the men at the stations, this practice being similar to that in vogue in New York. As regards the merits or otherwise of this system, much may be written and the subject is fully dealt with in a later chapter.

From this brief résumé it will be gathered that German fire control is planned on severely official lines, which to some degree no doubt stifles initiative on the part of the individual, but at the same time makes for that mechanical precision which is responsible for the fire risks being the lowest in the world.

The Berlin fire brigade was organized in 1851 and modernized in 1875, when the system of fire-fighting units was first brought into operation. Each unit comprised a trap, a manual and a water tank. Such was the organization at that time that, within ten minutes of a fire being reported, it was nominally possible to obtain the assistance of such an unit at any point throughout the city. As the brigade stands today it consists of a headquarters and five divisions, each division controlling five units. This force has to protect roughly an area of 15,000 acres with 27,800 buildings and a population of 2,123,000 souls. The officers of the department consist of one chief, two deputy chiefs, five divisional officers, fifteen assistant divisional officers in charge of "units" and two adjutants. By way of comparison Berlin has 25 officers to a brigade of 1,040 of all ranks, Vienna 7 to 468, Hamburg 12 to 512, London 5 to 1,400, and New York 63 to 4,996. Berlin authorities state that it should be borne in mind that it is not only the superior officer's management of the brigade and his greater technical education, but his general influence over the policy of fire prevention, his skilled assistance in the supervision and inspection of buildings, and the prestige which is conferred thereby upon fire control, which gives the department the standing it deserves as a highly important economic feature in municipal and national life.

Such recognition from so highly organized a body as the Berlin Municipality makes the writer hopeful that before long the status of the genus fireman will cease to be regarded as less important than that of the soldier or the sailor in the service of his country.

The policy of the Berlin fire department, as might be expected, has been towards the adoption of mechanical means of transport, and at the present moment most of the units are equipped with automobile appliances. Chief amongst these may be noticed a number of eighty-foot extension ladders, chemical engines, and steam pumps. There are also electrically propelled break-down cars for dealing with dangerous structures. The loose gear carried is of the

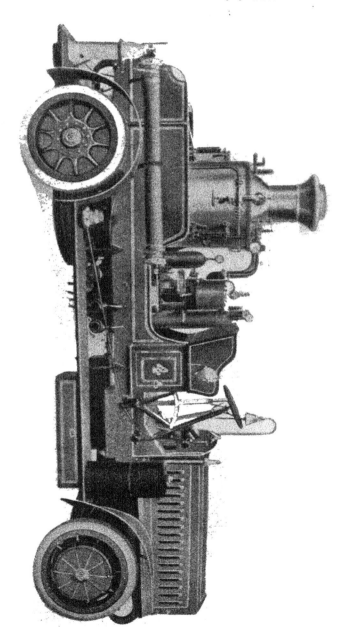

TYPE OF MODERN APPARATUS USED IN EUROPEAN CITIES

most extensive character, the men having special instruction in the use of smoke helmets, their familiarity in the employment of the same being second to none in the world. A feature is made of what may be termed "fire" tactics, or the topography of districts in the municipal area, enabling officers and men to fight fires to the best advantage. To this end a hand book is supplied, specially printed in order to be visible in a bad light, giving a tabular list of every thoroughfare and every hydrant or source of water supply in the city. But this compendium goes a step further, and in respect to particularly dangerous risks shows the most advantageous position to be occupied by individual engines. In the event of the apparatus designated being at work elsewhere, its place is taken by its relief.

Great attention is accorded to questions of fire prevention by the officers of the brigade, and systematic inspections are favoured, which are carried out by the brigade independently of, or in conjunction with, the Building Act Department, Factories Department or other sections of the police administration. The amount of inspection work done by the superior officers of the brigade in the last few years has been enormous, but its effect has also been very considerable in reducing the causes of fires. This supervision includes theatres and public buildings, factories, warehouses, department stores, hospitals, lunatic asylums and all buildings subject to special risks, such as electrical power houses and tanks for the storage of petrol and other explosives.

Those desiring to join the corps as officers must satisfy the authorities that they are physically sound, financially stable and possessed of first-class higher school certificates and military papers. They are then eligible to become ensigns, but in order to obtain commissions as officers they then must satisfy the authorities that they have passed the final examination as architects or civil engineers at a royal technical college, or taken the scientific courses at either a naval or a military engineering academy. Further, they must have been either commissioned officers in the army or navy, or at least hold rank in the reserve. In addition, they are required to place their financial position clearly before the Chief Officer, to undertake not to marry without the Chief's consent until they have been in the brigade at least one year, and to show that they have not only satisfactorily passed the ensign's course in the Berlin brigade, but also that of an ensign in at least one other brigade. Candidates must also possess a thorough grounding in electrical work and have a knowledge of the principles attending first aid. Finally they must be of good family.

From these details it will be seen that officers in the great German fire brigades must be men of exceptional ability, in fact, that the profession is practically closed except to those who, holding commissions in the army or navy reserve, are in private life architects or civil engineers, or belong to the engineering or artillery branches of the army or to the torpedo or gunnery branches of the navy. It seems superfluous to state that such credentials imply the acme of hard work and the height of scientific efficiency, yet the writer must be forgiven for hazarding the statement that the man trained in the university of hard knocks and who has gained his advancement from the ranks by shown ability to meet the emergencies of his calling, is in every way his equal for all practical purposes. The chief principles employed in fighting fires may be briefly summarized thus: Fight the flames at close quarters; always have a man in reserve on each branch armed with a life line and an axe for emergencies, and make use of all apparatus obtainable irrespective of immediate necessity. Comment upon these tactics is deferred till later in this article. It is, however, strictly enjoined that senior officers should not expose themselves to any unnecessary danger and should not under any circumstances penetrate to the heart of a fire or work inside buildings in danger of collapse.

There is a tendency observable to allow unimportant values to be destroyed if it is considered that their attack by water appears likely to cause greater damage than their worth justifies. Thus a roof or the contents of an attic, if situated in a high-class building, are generally allowed to burn out so that the floors below may not be injured by water. The rule of the brigade is to work upwards rather than downwards and a branch is rarely applied to a fire from surrounding elevated positions.

Regardless of the utility and great convenience of mechanically operated extension ladders, the brigade continues to give the closest attention to hook ladder and life line work. Every foreman and fireman must be thoroughly efficient in the operation of these two appliances, failing which he is compulsorily retired, and every fireman drills once a week at least with this apparatus during the whole period of his service.

Berlin possesses nearly seven hundred fire alarms, of which two hundred are public street alarms, directions thereto being fixed on every lamp-post, pillar-box and licensed kiosk adjacent to a crossing.

The regulations governing the department stores of Berlin are peculiarly comprehensive with the result that they are probably the best safeguarded in the world. Each shop of any magnitude has its private fire brigade, the watch room of which is centrally situated, and apparatus for any emergency is kept in constant readiness. Employees are specially trained as to the alarms, bell signals, appliances and those quarters to which they must proceed in the event of an alarm. Such signals are, (a) Quarters; (b) Return to Duty; (c) Clear Premises. The first signal can be "pulled" at any one of the private alarm points in the building; the second and third by a member of the private fire brigade alone, and then only from the watch room. Upon the first call sounding, those attached to the fire section proceed to the scene of the outbreak, which is marked upon a specially illuminated location chart, while those not similarly engaged are expected to remain at their posts under pain of instant dismissal. By this method, anything in the nature of a panic amongst customers is immediately checked. At the third call the personnel not at quarters is expected to pilot the clientele into the open by exits arranged according to departments. Meantime all wagons and carriages have been removed from the courtyards, areas and so forth by a special staff of porters, who likewise act upon prearranged signals. As on board ship, test alarms are frequently made to familiarize both staff and visitors with the mode of clearance, and in this connection it is of interest to note that in the event of what a police officer may deem to be the overcrowding of any store he has the power of stopping the entrance thereto, until such time as the congestion has ceased.

Theatre fire risks in Berlin are inconsiderable, thanks to the modernity of the majority of these structures, coupled with the stringency of the building regulations. The natural tendency to roominess observable in all public construction in Germany has also beneficially influenced the internal designs of places of amusement from a fire point of view.

In Berlin proper there are thirty-four theatres, music halls and circus buildings. The daily fire watches number thirty-six foremen and one hundred and nine men, about one-sixth of the brigade, or a full half of the men off duty, for it must be explained that the men forming this contingent are voluntarily recruited from those who, in their spare time, wish to make extra pay. It must, however, be borne in mind that this special service is compulsory as regards the brigade, the means of its supply being left to the Chief of the department.

The problems connected with safety in stage illumination appear to have been solved in a satisfactory manner. Effects of flames and fire are obtained by concentrating electric lights of considerable power and of the required colours upon pieces of silk which are suspended by one end and blown into position with a fluttering movement by electric fans and bellows. A duplication of the lighting system is also provided in most theatres, this being obviously of extreme value in cases of emergency, when otherwise the building would be plunged in darkness.

An example of excellence in, theatre construction is

afforded by the Schiller theatre, with seating accommodation for 1,460 persons, which it is estimated can be emptied in less than one minute. There is only one gallery of small size, the rest of the house being given over to what corresponds to stalls and pit in European theatres, or in American phraseology, "orchestra chairs."

To understand the situation of the fire service in Hamburg, it is necessary to appreciate that this is a city, which in the main is a port with enormous warehouse values, both within the dutiable area and in the "free port"; that it further has a large city or office district, a retail business section and finally extensive residential suburbs of varying descriptions. The business portion of the city is intersected by a large number of waterways, which, whilst providing the most valuable auxiliary of an ample and accessible water supply in some of the more dangerous districts, at the same time create considerable difficulty for intercommunication and the concentration of the brigade in force. Roughly, the population of Hamburg amounts to 900,000, the number of buildings approximating 31,000.

The main fire risk is naturally centred in the warehouse area and more especially in the "free port," where, owing to the shortsightedness of those responsible to the harbour board for the dock equipment, constructed in the early eighties of the last century, buildings were erected with all vertical and horizontal metal supports entirely unprotected and in many cases formed of light lattice work girders, which are peculiarly liable to collapse when subjected to great heat.

In the newer warehouses all this has been remedied and the improvements introduced include the use of fire resisting materials to protect supports, the substitution of ordinary flooring by reinforced concrete laid at such an angle as to insure the speedy and easy drainage of water into scuppers, thus avoiding unnecessary damage therefrom in the event of fire, and the absolute insulation of all elevator shafts and staircases from the rest of the building.

As to the development of the fire department, its history is short considering the lesson that should have been learned from the destructive conflagration of 1842. Not until 1869 was a professional brigade formed, and then it consisted only of the ridiculously inadequate number of forty-eight men under a chief officer. This in turn was assisted by twelve hundred volunteers, the apparatus at their joint command comprising four steam fire pumps and one hundred and nine manuals. In 1878 the force was reconstituted and today it consists of a chief officer, twelve assistants, six warrant officers, forty-three foremen, twenty-nine engineers and four hundred and twenty-two firemen, or, together with supplementary staff such as telegraphists and electricians, a total of nearly five hundred and fifty men. There are ten fire stations and, as in Berlin, the hours of duty are forty-eight on to twenty-four off. Some idea of the brigade's activity may be gleaned from the fact that on the yearly average it attends seventy fires of first importance and one thousand of lesser importance, while false alarms total the huge number of nearly five hundred. It would be of interest to know to what the latter remarkable figure is attributable.

The equipment of the brigade is excellent. Amongst other apparatus may be noticed twenty-five steam fire pumps, seven chemical engines, ten eighty-foot extension ladders and no less than seventeen large fire floats.

Considering the strength of its personnel, the area of the city and the property to be protected, it is no exaggeration to state that few fire brigades can show so large a proportion of mechanically equipped apparatus, which, in itself, speaks volumes for the enterprise of the responsible authorities. The administration of the force is in the hands of a special civic commission, formed on comparatively independent lines and representing the various interests at stake, both financial and technical. It consists of a Senator who acts as Chairman, a lawyer from the Senate, three municipal councillors, two municipal fire insurance officials, and an official from the city's water works. The cost of

the brigade amounts annually to $450,000, of which $240,-
000 is raised by a special rate upon house property, $50,000
by stamp duties on fire insurance policies, while the remainder is provided by the authorities out of their general
funds.

An interesting feature is the position occupied by the
chimney sweep, that humble individual whose services seldom receive recognition of any sort from the community,
yet upon whose thoroughness depends the safety of property
and persons untold.   In Hamburg the genus sweep is under
fire brigade control, and no one can start in that business
without first passing a stringent examination.   It is compulsory to have all chimneys cleaned at regular intervals,
and in the event of negligence, both sweep and proprietor
of the premises at fault are heavily fined.

Generally, as regards fire risks, the Hamburg municipality has framed special by-laws along much the same lines as
those existing in Berlin, and the protection thus afforded
is both ample and adequate.

Though the town of Hanover is small, its population amounting only to 272,000, anyone visiting its
brigade cannot but be struck by the fact that it
is no ordinary organization, but rather one of exceptional
excellence and which on that account can afford to be
compared with any in Europe.   Needless to say, any great
expenditure on apparatus cannot be expected from such
a small community, but the district covered possesses a
dangerous manufacturing section and includes some factories of great size.   Hence, to meet the needs of the situation it has been necessary to provide a department which, if
confined to its regular duties, would scarcely find sufficient
employment.   But an economical solution of the problem
was found by according to the brigade and its officers
additional municipal functions other than those of the fire
service, and to this end both officers and men have been
trained for other special duties.   At the same time it was
wisely determined that the apparatus, though limited in
quantity should be the best obtainable in quality, and that

the salaries of all concerned should be upon as liberal a scale as possible.

Thus, the brigade acts as the ambulance department of Hanover, in itself a work of considerable utility. Unlike other departments, which possess a first-aid section, in this case the corps undertakes the transport of infectious cases and the like to hospital, which though open naturally to serious objection on account of the possibility of the spread of disease through this agency, is none the less a service that in a small town can be carried on with the minimum of risk, when every man concerned is under the closest medical supervision. In addition, the chief officer of this fire department is also, *ipso facto,* the administrative head of the municipal scavenging and dust destructor service, which incidentally has considerable bearing upon fire prevention. Though no doubt a certain sympathy must be felt for scientific fire-fighters, who are expected to employ a portion of their time in such uncongenial occupations as taking diphtheria patients to hospital or acting as scavengers, yet as the municipality urge, they can only afford to pay for a brigade in which the rank and file can be otherwise employed, and it would seem better to have a fire force at even that price than possess none at all. And it must also be remembered that Germany is a free country and that there is no compulsion to serve, at any rate in the Hanover fire department.

The present constitution of the brigade is as follows: four superior officers, an inspector of telegraphs, a superintendent of ambulance work, seventeen foremen, eighty-six firemen, six telegraph clerks, and twelve coachmen, or one hundred and twenty-seven of all ranks. There are three fire stations and approximately 13,800 buildings to be protected. The principal equipment consists of three steam motor-propelled pumps, three eighty-foot extension ladders, four motor chemical engines and seven traps. These latter are extremely useful appliances, carrying hook and scaling ladders, a quantity of hose, life lines and all those minor appliances that at fires often spell so much at the commencement of an outbreak. Three motor ambulances also merit mention, and all municipal telegraphy and electric wiring for bell and signal purposes being under the brigade's control, there are special motor trolleys for that branch of the department. The corps is equipped with forty-five street alarm call boxes in public thoroughfares, and twenty-two in private or municipal buildings. On the average, the annual number of fires attended amounts to two hundred and eighty-two, of which twenty-one rank as of major importance, twenty-eight are medium and seventy-eight are chimney fires. The ambulance section roughly answers 4,500 calls per annum of which no less than six hundred may be docketed as infectious. Hence it speaks volumes for the medical precautions adopted that rarely, if ever, a fireman is temporarily incapacitated or permanently injured from this duty.

As indicated, though a mere enumeration of personnel scarcely serves to emphasize sufficiently the point, this small force is no ordinary one and under its former fire chief, Herr Reichel, now in command in Berlin, it can lay claim to having taken the initiative in motor traction as applied to fire engines, certainly in Germany, if not in the entire world.

Today there would be nothing in a fire brigade ordering self-propelled appliances, rather would they be remarkable if they did not. But it is worthy of more than passing comment that as long ago as 1901, Herr Reichel was able to exhibit at the Berlin International Fire Exhibition, a complete fire service unit for a district station, comprising a motor steam fire engine, an automobile trap, and a self-propelled chemical engine, which, working as an unit, time has proved to be eminently economical. The unit in question, after an experimental trial of three months, entered the regular service of the Hanover force and is still doing excellent work even today.

As regards water supply this is ample, the pressure off the mains averaging forty-five pounds.

Before closing the brief account of this most enterprising small brigade a few words must be added concerning the actual methods employed in the ambulance service.   On an alarm sounding an ambulance starts away at once in charge of a coachman and four firemen.   In infectious cases the men have instructions to handle the sufferer as little as possible, and at the end of the journey both attendants and coach are thoroughly fumigated.   This system is also used for the removal of dangerous persons and lunatics, thus constituting a valuable auxiliary to the local police, hospitals, and lunatic asylums.   Finally, this branch of the brigade during the summer months is charged with the manufacture of ice, which is sold at cost price to those in a position to pay for it, but is supplied free to the poor in case of illness or other necessity.

In fact, the town of Hanover can lay claim to the proud boast that first of all the cities in the world it has recognized the science of fire-fighting to the extent of founding a lectureship on "Fire Control," the chair of which is located at the Royal Technical College of Hanover, which now ranks as a national university.   The first lecturer, "docent," was that Herr Reichel, of whom mention has already been made.

From a perusal of the foregoing pages the reader will have recognized that the outstanding feature of German fire brigade organization, as evidenced by that of its most important centres, is the large part played by a semi-military handling of the subject, coupled with that thoroughness of technique and design, which is distinctive of the Teuton character.   But this must not be taken to mean that in the opinion of the author, nothing is beyond criticism or above discussion.   In the first place, as must always happen in countries where class distinctions are rigid and the private soldier cannot in all truth be said to carry the Field Marshal's baton in his pocket, there is that tendency to assume that mere theoretical training is sufficient to equip an individual satisfactorily to fight so insidious an enemy as "fire."   It is the humble opinion of the writer that this theory is erroneous.   The individual may be provided with the most extensive scientific panoply of degrees and diplomas regarding the arithmetic progression of combustion under certain conditions, he may be able to work out by trigonometry the angle of a water delivery from a pump to a window many feet from the ground, and he may be an expert at assessing the nozzle pressure necessary successfully to circumvent an outbreak, before the latter has reached serious proportions. This in theory. But what of the practice. Every sailor knows that it is a matter of no great difficulty to ascertain in a class room the position of an imaginary ship upon an imaginary ocean, with the assistance of an imaginary sextant and the ordinary aids to navigation. Everything is at hand to make his task an easy one, even to that of such adjuncts as light, warmth and stability. But place that same individual on board a real ship upon a real ocean, in a small ill-lighted deck-house, with a chart pinned down on a swaying, uneven surface, and ask him to work out the same set of figures or the same problem, and he may be forgiven if he fails hopelessly. So is it in all appertaining to this science of fire-fighting. With all the technical knowledge in the world and nothing else behind it, it would be ludicrous to expect any person successfully to cope with so crafty an enemy as the flames, or at any rate as competently to obtain their mastery as one trained actually upon the field of experience.

In this connection, also, without wishing to appear hypercritical, it seems doubtful whether the Berlin practice of preventing senior officers from taking an active part in the actual fire-fighting is either wise or desirable. True, a general on a battlefield is expected to direct operations from a point of as much safety as is consistent with his duties, but in the case of a fire chief, it should be remembered that each fire must be fought on its particular merits. There has been no survey of the ground previously, there has been no active intelligence department to warn the attacking force of what particular line of development may be expected; all that the fire chief knows is the bare fact that an outbreak has occurred at such and such a place and that the locality is a dangerous one or *vice versa*. Hence, in order to satisfy himself as to the true state of affairs it is imperative that he should judge for himself by personal observation as to the possible chances of a spread of the flames and the best method to fight the same.

Further, another feature of the Berlin fire department seems to demand special criticism, namely the custom of allowing a fire to burn itself out if situated at the top of a building, the other contents of which would be damaged by water attack. No doubt this may be essayed and essayed safely in a fire-proof building, separated from its neighbors by a certain distance, and when a sufficient portion of the fire department is concentrated on the scene and can remain there for any emergency. But time must be allowed for the said fire to burn out, and the force detailed to watch it may meantime be urgently wanted elsewhere. To leave it unwatched would, of course, be suicidal. Hence, such tactics must be regarded as hazardous, and much better were it that the insurance companies should suffer for a minimum of loss than be obliged to meet the demands of a really serious conflagration, the possibility of which is always present under such conditions. These are a few of the thoughts which arise in the mind of any trained practical fire-fighter. It is the theoretician who sees in the vicarious strategy outlined above a better method of overcoming a wily enemy than the old style of coming to grips at once and fighting to a finish.

For the rest, the German fire departments have much to recommend them as models to the world, not the least important factor in their organization being the prestige attaching to fire-fighting as a science, and to the honourable position occupied by officers and men in the estimation of the public

CHAPTER VII

THE dominant feature of the Austrian fire department is the high degree of excellence attained by purely voluntary corps, which owe their development in a great measure to the system of federation introduced as long ago as 1869. This organization was extended in 1885 under the name of the Austrian Fire Brigade Board, comprising delegates from the provincial brigades under a president and two vice-presidents. In 1900 this Board received recognition from the Crown, became known as the Austrian Imperial Fire Brigades Association and obtained an annual subvention from the Government. Today this federation numbers more than 9,500 brigades representing practically the entire Austrian service. Some idea of the magnitude of this force can be gathered from the amount of the apparatus involved, namely, over 200 steam fire engines and over 13,000 manuals. From time to time this Association appoints technical commissions to examine all questions connected with the scientific aspect of fire control. Further, courses of study are specially designed to familiarize officers and non-commissioned officers with the theoretical problems involved in fire-fighting. Particular attention is directed to the inspection of local brigades, and efforts are made to secure uniformity in their organization. According to the principles adopted by the Imperial Association, every volunteer corps must be composed of two sections, each properly equipped for fire extinguishing and life saving work. Thus these brigades can be used not only for fighting fires, but may equally be called upon to render assistance in the event of accidents of any nature. Such a system is no doubt of public benefit in rural communities, but would clearly be impracticable in large towns unless all municipal forces were under the control of the same executive.

A matter of enormous importance in the economics of the Austrian fire service is the fact that the law of the country requires all insurance corporations and companies, trading on Austrian territory, to contribute about two per cent. of their total gross premium income on the risk taken in Austrian territory, for the specific purpose of assisting in the upkeep of the fire brigades and towards the firemen's widows and orphans' fund. This law affects all companies, irrespective of their nationality. It must be here emphasized that the foregoing remarks apply only to Austrian territory proper, Hungary possessing a distinct and separate organization of its own. The Hungarian Fire Brigades Union consists of 1,325 units out of a total of nearly 9,000 corps, sufficient evidence of the fact that it has not won the same popular interest. Its executive serves as the board of experts, to which the Minister of the Interior applies when technical questions have to be dealt with. An annual course of instruction is arranged by the Union lasting three weeks, and no officer apparently can attain chief officer's rank in a Union brigade without having passed this test and obtained a certificate. As to the Union's general work, it has systematized all questions of uniform and badges of rank, it has created a long-service medal and has issued clear instructions for competitions and a guide for the testing of fire extinguishing appliances. Doubtless this list of ordinances is possessed of local value, but to the scientific mind it seems strange that questions of technical import have not received more attention from such an association. True, a uniform coupling is used throughout the country, and there is a standard manual fire engine, but as non-union brigades possess these appliances, it may be pre-

sumed that their adoption has merely been a matter of convenience.

The city of Vienna, as regards fire protection, is dependent upon a municipally paid professional brigade assisted by volunteer suburban corps under the control of brigade headquarters. Eight officers, five civilians and 475 men form the personnel of the former, located in 15 stations, with two special watches in public buildings. The officers consist of the Commandant, a Chief Inspector and six subordinates, all of whom are housed at the central fire station. Of the rank and file, 8 are drill sergeants, 40 telegraph clerks, 53 foremen, 22 engineers, while 248 comprise the actual fire-fighting force. In addition 24 telegraph clerks and engineers are detailed for duty with the volunteer suburban brigades, the remainder of the force, numbering 78, being coachmen.

Numerically such a fighting strength for the fire protection of a city of the size of Vienna would seem hopelessly inadequate, but in this connection a word must be said for the building regulations enforced by the municipality, which greatly diminish fire risks owing to their far-sighted efficacy.

The apparatus of the brigade is adequate to its needs, perhaps the most distinctive feature being the chemical engines, in connection with which are operated 80-foot mechanical extension ladders. Their crew consists of an officer and five men, additional gear carried comprising 3 hook ladders, a hose reel, a hand engine, a smoke helmet, a jumping sheet, an ambulance chest, toolbox, torches, and so forth. On duty the firemen wear uniforms of white canvas, which scarcely seem appropriate, considering the nature of the work they are called upon to do. Generally speaking, it would appear from the data obtainable that there is a tendency to overload the men with gear, and that some of the heavier apparatus is insufficiently supplied with personnel effectively to operate it.

The Suburban Volunteer brigades turn out to fires in their own districts, but may be called upon to assist in the event of a serious outbreak in the city. Their equipment is very similar to the municipal brigade and, since the men are volunteers, and as such enthusiasts, they take a pride in keeping as up-to-date as possible in all matters pertaining to their apparatus.

Vienna is particularly fortunate as regards its water supply, which is ample for the requirements of the brigade. There are 3,620 hydrants, with an average nozzle-pressure of from 75 to 90 pounds, so that the use of the steam fire engine is rarely necessary, doing away with much of the cumbersome apparatus found in other continental cities.

The Hofburg Theatre is generally considered one of the finest in the world, but judging from the following report from the Journal of the British Fire Prevention Committee it would seem that, in common with most other similar structures in the European capitals, desire for architectural magnificence has outweighed the less artistic essentials upon which its fire safety depends. "It is subject to a considerable risk of fire through antiquated electric installation. The switch room on the stage is one of the most dangerous the members of the party have seen. Besides being dangerous electrically, it is highly inflammable, and lined with match-boarding. There was much unnecessary match-boarding and woodwork in the theatre. It appeared curious that a building such as the Vienna Hofburg Theatre, on which such an immense sum of money had been spent, should contain defects so palpable that they were inexcusable. The staircase from the stage to the mezzanine was very antiquated, and capable of much improvement. The exits, however, seemed ample." The theatre has its own fire staff, and below the gridiron is fitted with a species of sprinkler operated from the stage level. Beyond this precaution, apparently nothing is done to ensure the safety of the flies or the scenery dock. There is, of course, an iron curtain between the stage and the auditorium, though the front of the house is seemingly left unprotected. At all theatres in Vienna an evening watch is posted and the fire apparatus is examined prior to, and after, each performance.

Though an Ambulance Service can scarcely be considered an integral portion of a fire department, yet in Vienna the two organizations are so combined as to be almost inseparable. Formed consequent upon the Ring Theatre fire of 1871 by Baron Mundi, the Vienna Volunteer Ambulance Society has as its object the creation of a civil ambulance service to render aid on occasions of great emergency, such as conflagrations, railway accidents, floods and the like. It consists of three departments, the first detailed for fire service, the second for flood service, and the third for first-aid service. The fire service comprises several of the Vienna suburban volunteer fire brigades, four hundred of the men of these brigades being organized to do duty for this purpose outside the metropolitan area, if necessary. The flood service comprises 149 men from the leading rowing clubs, and has its own pumps, pontoons and food distributing vehicles, thus acting to some degree as a substitute for a regular river fire department. The first-aid service comprises 14 paid doctors, 325 voluntary doctors, 60 medical students, 3 ambulance superintendents, 12 ambulance orderlies and 6 coachmen. In the administrative building of the society there are waiting-rooms, duty rooms, an accident ward, operating theatre and watchhouse, the latter specially equipped with telephones for communication with the fire department, police and other authorities. For railway accidents the radius of action is 300 miles, while in the event of a conflagration or great disaster the society can count immediately upon the services of 50 doctors and 200 volunteer ambulance orderlies equipped with 26 ambulances, 250 stretchers and a large quantity of minor appliances. This forms a valuable auxiliary to the fire department, which can always rely upon its immediate coöperation.

Some consideration must now be given to the Buda Pesth fire brigade, which is likewise a combination of professional and volunteer forces. The staff of the professional brigade consists of a Chief Officer, an Inspector, a Senior and two

Junior Adjutants, 23 warrant officers, 3 engineers, 15 foremen, 175 firemen and sufficient coachmen to drive the
horsed appliances. Amongst the apparatus may be noticed
16 fire engines, 22 manual engines, and a supply of hose
wagons and extension ladders. Headquarters and sub-
stations are connected by private telephones. There are
149 fire alarms distributed throughout the city, which number seems inadequate. Since the publication of these data,
it is understood that arrangements have been made to re-
equip the force, but necessarily this operation will cover
some time. On the other hand, the Volunteer Brigade is a
model of its kind, possesses an independent constitution and
comprises some 80 members. It is capitalized to the extent
of $40,000, and receives in addition a special annual sub-
sidy from the municipality. Though legally an entirely
self-governing institution the corps voluntarily puts itself
under the command of the Chief Officer of the municipal
brigade. Their equipment is housed. together, since that
operated by the volunteers is bought and maintained by the
city. The professional head of the department has at his
daily disposal ten men, who do duty every night and render
service if called upon. Owing to the fact that the fire risks
in Buda Pesth are regarded as considerable, it has been
found necessary to augment these two services by essentially private organizations of factory fire brigades. These
number 44 all told, total 1,600 men, and have a mutual
understanding whereby the members of any one factory
assist others in case of need.

In criticizing the fire department and equipment of a
town such as this, it must be remembered that it would be
expecting too much to demand the finished organization and
up-to-date resources of a city such as New York. When it
is considered that only latterly has fire control come to be
regarded as worthy of more than passing attention, it
speaks volumes for the enterprise of a municipality situated
so far east and peopled by a race so temperamental
as the Hungarians, to have evolved so efficient a service. This comment is made necessary because, since com-

BUDA-PESTH VOLUNTEER FIRE BRIGADE.   BENZINE MOTOR FIRE ENGINE.

parisons are odious but constantly instituted, it may be imagined that such a statement of facts implies discredit.

The following condensed account of the burning of the "Parisian Store" in Buda Pesth on August 24, 1903, though ancient history, still possesses considerable interest. On the ground and mezzanine floors of the building were business premises, while the other four stories comprising the house were given over to residential apartments. An open courtyard in the centre of the block provided light and air to the residential portion. The proprietor of the business premises, wishing to increase his accommodation, had rented the mezzanine floor of the two adjoining blocks, cutting large openings in the party wall. In addition he roofed over the open court at the floor level above the mezzanine, closing the doors on the ground and mezzanine floors leading to both front and back staircases and blocking the windows facing the business premises. The store premises were stocked with drygoods.

At about 7 P.M. smoke was seen issuing through the partition separating the business from the main street entrance of the residential portion. It is alleged that the outbreak was due to an electric short circuit, but more probably it originated among some of the inflammable goods in the store. The fire spread rapidly, volumes of smoke cutting off the egress of the tenants. Shortly the whole of the business portion of the building was involved, and the flames entered the residential part through the glass roof over the central court. Thus the tenants had no other means of escape except the windows overlooking the street, the door of the back staircase having meantime become involved in the general conflagration. Before the arrival of the brigade three persons had jumped from windows and lost their lives. By the time that the brigade had arrived upon the scene the fire had obtained so firm a hold that the fire escapes and jumping sheets could not be employed to proper advantage, with the result that 26 other persons jumped, of whom 9 lost their lives, 16 were seriously injured and one was unharmed. Owing to the openings in the party walls of the mezzanine, the fire spread to the adjoining block, narrowly avoiding a very much larger area of damage. The moral of such a calamity is obvious. When tenements are over business premises, every constructional means should be adopted to ensure the safety of the residents. In this connection the municipality itself should see to it that in all new buildings attention is paid to fire risks, and also that no trade or business of a dangerous nature should be carried on in any inhabited dwelling. This is of especial importance in these days, when the employment of celluloid in various forms has come into such common use.

Since the development of Rome from 1870 onwards, combustible materials in building construction have been practically prohibited. In buildings prior to 1870 wood could be primarily found only in roofs and floors. The wooden staircase in Rome is an exception, and in structures both old and new, a substantial vaulted fire-resisting floor separates the ground floor from all other parts of the building. Thus all shops on the street level are effectively isolated from tenements above. The number of factories and workshops in Rome is small and is limited to a few steam mills; consequently up to 1894 the fire brigade was composed of municipal workers, who took it in turns to man the stations and to act as theatre watchmen. Since that year the force has been reorganized, being 200 strong, of whom 140 are firemen, 50 belong to a special reserve and ten are officers. The municipality pays the entire expenses of the brigade, amounting to about $12,500 per annum.

There are in all seven stations connected by telephonic communication, and an alarm system of roughly a hundred points. As regards water supply, there are 350 hydrants exclusively for fire purposes, together with some 3,000 others, which can be brought into use if necessary. It is estimated that the number of fires per annum amount approximately to 270, of which on an average 216 may be listed as "petty," the damage incurred being in each case under $200. Since the population of Rome aggregates half a million, it will be seen that the incidence of fires per thousand inhabitants works out at only 1.8.   The total average fire damage annually reaches $50,000.   In case of necessity, following the usual continental procedure, the brigade renders assistance at disasters other than fires.   As far as apparatus is concerned there is little to demand attention, the equipment for the most part being somewhat antiquated.

No better illustration of the divergency in Italian temperament could be exemplified than the organization of the Milan fire department and that of Rome.   The northern capital is keenly alive to fire risks, and with that enterprise which distinguishes the Piedmontese it has left no stone unturned to keep its equipment at a high level of excellence. By the decree of the Viceroy, Eugene Napoleon, the brigade was first organized in 1811, and consisted of 2 officers and 81 men, who were exempt from military service, but were under military discipline.   This jurisdiction was not removed till 1859.   A great fire, which occurred in 1871, shewed the necessity for the augmentation of the force, and in the following year 100 members were added, divided into two sections of 50 firemen each.   The first was formed of regular firemen posted at the stations; the second of workmen who were obliged to undergo a periodical instruction, attend fires, and undertake patrol duty in the theatres.

In 1905 the corps was modernized, and the present personnel comprises eight superior officers with 240 rank and file.   The superintendence of the equipment is delegated to a chief engineer assisted by a motor expert.

Included amongst the appliances are 86 manuals and 9 steam fire engines, 5 motor-driven pumps and 9 extension ladders.   The use of the chemical engine is general, and a large supply of smoke helmets is included in the apparatus.   There are seven stations with direct telephonic communication, each being specially connected with the municipal offices, the police, the military and the theatres.   On an average per annum there are 785 alarms, of which 16 are

serious, 52 of less importance and 659 of slight consequence. False alarms are inconsiderable. In addition the
brigade renders first aid, being provided with special ambulances for that purpose, while it assists also in the de-
molishing of dangerous structures.

The Scala Theatre, Milan, is world renowned on account
of its vast size, being third in seating capacity of all such
structures. It is subject to the supervision of the theatre
committee, but being a building of considerable antiquity
and very inferior in fabric it can only serve as an example
of how a theatre may escape destruction by fire, regardless
of the fact that the most elementary rules of constructional
equipment have been disregarded. Hence great credit must
be accorded to the Theatre Committee in its efforts to obtain small improvements, whilst not having the required
powers for the drastic action necessary. The hydrants in
the building have a nozzle pressure of about 40 pounds at
the stage level, and are so arranged that the upper floors
may be served through their being coupled to steam fire
pumps.

Another feature in Milan is also worthy of note. As in
many continental countries the Government of Italy has
taken over control of all pawnshops, and has organized
them into a State Department, known as the Mont de
Piété, which comprises, besides the actual loan office, a
credit bank and a safe deposit. For this purpose the Municipality of Milan has constructed a special fireproof building, which of its kind is a model. Of reinforced concrete,
the floors of the galleries are of iron with cages of steel
wire for the storage of goods in pawn. There is a special
watch station on the top of the highest portion of the building, connected direct to fire headquarters, and a special
patrol is kept constantly on duty. Incidentally there are
some 60,000 depositors per annum, and nearly 65 per cent.
of the goods pawned are under the value of $4; the total
value of pledges in one year reach the enormous sum of
$2,300,000, a sufficient indication of the use made of this
institution. By a Government regulation, when a reserve

fund of $50,000 has been accumulated, the profit goes to municipal charities, so that the money of the needy may be said to supply in part their own necessities.

The Florentine fire department is the best volunteer organization of the kind which can be found in Italy. It is commanded by a military officer, specially selected from the army for this purpose, and paid by the municipality, which also provides the equipment and fire station. Otherwise it is officered and manned by volunteers, numbering about 130 officers and men. Their apparatus consists of 4 steam fire engines, a salvage and dangerous structure trap, which is in itself something of a novelty, and three extension ladders. Florence has about 160 fires annually. Since the water supply is not altogether satisfactory, and hydrants are not to be found in all streets, special engines are used capable of drawing water at a distance of over 300 feet. When the pressure is too small, pumps are used in tandem. The average power from the mains is about 40 pounds to the square inch, which is sufficient for the services it is called upon to perform.

Needless to say, the part played by the fire brigade in Venice is one which, in some of its aspects, is unique. Naturally in a city with canals as highroads, the question of transportation differs materially from that in other towns. The corps forms an integral portion of the "Vigili," or municipal watchmen, who preserve order and generally render assistance to the community.

Thus in the event of a serious conflagration, the police section of the "Vigili" augment the fire section and *vice versa*. Each division has a commander and its own staff, both being under the supervision of a military officer specially appointed by the municipality. The rank and file of the fire department number 71, and are distributed in 6 companies of varying strength. Their apparatus is naturally designed for water transport, and consists of one large modern petrol propelled float, one large old type steam float, two 35-foot steam launches, and several small petrol motor boats, which are used as first-aid appliances.

Manual engines, ladders, and so forth, are carried in a large fleet of swift gondolas. Fire escape work is done with Roman ladders, which are usually planted on two gondolas slung together barge form, or, if the depth of the canal permits, the lower length is bedded in the canal bottom.

Owing to the substantial character of the older buildings, and also of the modern residential and business structures, the fire hazards are primarily those in the dock area, with its numerous sheds and small warehouses of a highly inflammable character. There are also some large industrial works in which the fire risks are equally great. The number of fires annually is comparatively small, averaging 125, and it is rare that more than one or two can be classified as serious. Roughly the fire loss per annum is $50,000, or about $400 per fire.

Generally speaking, a considerable awakening of interest in questions relating to fire control is manifest in Italy, King Victor being something of an enthusiast in that respect. It is a mistake, however, to suppose, as is advanced by some technical writers, that Italy is more immune from the fire peril than other countries, because of its climate. Facts speak for themselves, and the fire risks in New York are nearly as great in mid-summer as in the depths of winter.

Italy's geographical neighbour, Switzerland, possesses a fire service run practically on national lines, that of Zurich supplying an excellent example. This is a compulsory militia brigade under the control of the Chief of Police, who is also Chairman of a committee of nine charged with the protection of the town from fire. Zurich covers about 12,000 acres, 1,500 of which are built over with some 15,000 houses, the whole of the buildings being subject to the local building regulations and the "State Insurance Association's" rules in which they are compulsorily insured. Every male inhabitant of the town is compelled to do some service for the prevention of, or protection against, fire, from the age of 20 to 50, which duty may be fulfilled by active service or, in the case of an able-bodied citizen, who is found unsuitable for such service, by the payment of a tax. This impost is fixed upon the basis of his income, though certain citizens are, *ipso facto,* exempt from active fire duty. The fire brigade comprises 15 companies of 120 men each, the officers being appointed by the Municipal Committee. Only men, who are personally enthusiastic, and who are possessed of good physique, are selected, and are preferably recruited from the building or allied trades. Absence from drills is regarded as a serious offense, being punishable by a fine alternatively with imprisonment. The city insures the whole of the brigade against accidents and illness with the Swiss Fire Brigade Union, and also provides a fund for families in cases of death of firemen on duty.

Each company has three sections: a fire service section, a life saving section and a police section, the latter being utilized for keeping the ground free and attending to salvage. Further, each company is supposed, as a rule, to be able to deal with any fire in its own district, and it is only in the case of a very serious outbreak that additional companies are requested. Thus there is a system of decentralization and independence of action in this force not often met with elsewhere, which, applied to a large area, would be unworkable. Firemen receive 20 cents for each drill of two hours, while for fires they receive 40 cents for two hours and ten cents for each additional hour. This would appear to provide an incentive to unscrupulous firemen, though probably such are non-existent in Zurich, to prolong the life of a fire in accordance with the demands of their purse. The official regulations also state that refreshments are provided, though in this connection it is not clear whether before, during, or after a blaze! ! !

An extensive telephone service is at the disposal of the brigade, but since all the personnel are not connected with the system, the alarm is mainly given by horns blown by those who have telephones in their homes. One may be forgiven for imagining that under such circumstances this number cannot be very great. By law, the telephone service is free for alarms, and is at the disposal of anyone for that purpose.

A company comprises 1 chief officer, 1 second officer, 1 doctor, 2 ambulance men and 6 orderlies as staff in charge, supplemented by: for the fire service, 1 lieutenant and 40 men, for the life-saving section, the same, and for the police section, 1 lieutenant and 20 men. The full force of all companies is about 2,300 of all ranks. The apparatus is simple in nature, consisting mainly of hose reels and ladder trucks, housed in corrugated iron sheds to which the firemen all have keys. This simplicity of equipment is only made possible by an excellent service of hydrants, of which the city has 2,895, with a nozzle-pressure of from 60 to 120 pounds. This represents a great advantage over the pressures to be found in most other continental cities, and is attributable to the fact that the water supply comes from the mountains.

The fire control service is organized on most elaborate lines, owing to the fact that the building regulations and state fire insurance are practically in the same hands. All fresh construction and even alteration is subject to a Cantonial Building Act, and it is the duty of the building department to carry out the law. Three members of the Town Council form a committee to grant or refuse licenses for new buildings or alterations to old ones, and in this duty they are assisted by technical advisers, namely, the city architect and a number of architectural assistants and surveyors. In the case of a license being refused appeal may be made to the Town Council *in plenum,* and finally to the Cantonial government. Amongst the regulations is the stringent inspection and cleanliness of chimneys, and the officials are, *ipso facto,* liable to prosecution in case of an outbreak of fire, if it can be shown that they were guilty of neglecting their duty.

Such regulations speak volumes for the intelligence of the city fathers of this Swiss town, and are evidence of the realization by the municipality of the necessity for efficient fire control.

The principles underlying the organization of the Lucerne fire department are very similar to those governing Zurich, with the difference that there is not so much decentralization, and that the force is more homogeneous in character. It possesses, however, one feature which is probably unique. Attached to the life-saving section of the corps is a technical division composed of experts drawn from such industrial undertakings as the Municipal Electrical Supply Company, the Telephone Company, the Tramway Company, the gasworks and the waterworks. The officer in command of this section is a civil engineer on the regular staff of the brigade, whose duty it is to advise the commanding officer on all technical points.

All these divisions and sub-divisions must tend toward some confusion in practice, but at the same time the fire chief has ever at his disposal a fund of highly scientific information upon which to draw in case of need.

It may be emphasized, however, that the actual exigencies of fire-fighting under the conditions common to fires of any magnitude can not permit of any fire chief accepting or soliciting advice from any quarter. He must be sufficient unto himself in the moment of action, though naturally he may have imbibed much useful knowledge from such sources during official discussions. Anything that in the smallest degree tends to diminish the initiative of the fire chief must be disadvantageous to a proper grasp of his complex duties, and it is to be feared in this case that in a multitude of counsel is confusion. This is penned in no critical spirit, but rather as embodying the experience of a practiced fire fighter.

CHAPTER VIII

THE TRADE OF ARSON

IT is calculated that incendiarism for the purpose of obtaining insurance money is responsible for the destruction annually in New York alone of four million dollars' worth of property. This represents a daily loss of $10,000, or more than the yearly pay of a Major General in the United States Army. Needless to say this criminal practice is not confined to New York, every large town in America suffers in a greater or less degree from the attentions of the genus "firebug." Now for this state of affairs it is impossible wholly to acquit the great insurance companies, for latterly it has become usual to accept fire risks of considerable value without instituting the searching enquiries, which are a *sine qua non* for the completion of business in Europe and elsewhere. Of course, cases of arson do now and again occur in any community, but that a gang of criminals should find it both easy and profitable to carry on incendiarism as a regular calling seems almost incredible and bespeaks a species of toleration which is scarcely to the credit of the community. Quite apart from danger to public property and unnecessary loss to insurance companies stands out another point in the most vivid of relief, namely the dire peril to human life, of which these fiends take no account.

This murderous trade appears to be peculiarly lucrative, and judging from statistics it offers little risk to the perpetrators of discovery and punishment. In addition, also, it requires no stock in trade, such for instance as is necessary to the forger, it demands no courage such as characterizes and lends an air of romance to the train bandit and most assuredly it makes no great call upon mental ingenuity such as marks the operations of the bank swindler. Hence the "firebug" may without doubt be classed as belonging to the lowest and most degraded portion of the criminal population. Not that necessarily the votaries of this occupation lack a certain amount of spurious education. On the contrary they are drawn from all grades of society, the less educated being as a rule the tools employed to do the actual burning. In this category must also be included those misguided individuals, who finding themselves in financial difficulties regard a fire as the simplest method of retrieving their shattered fortunes. Frequently such people employ the services of the professional firebug and share the proceeds. Thus "fire making" has become a regularly accepted calling, which it is most urgent should be stamped out in its entirety once and for always.

Were additional evidence of the accuracy of these statements needed it is surely supplied by the following curious circumstances. During the spring, fires in the fur trade are prevalent, while hat and cap fires usually occur in the summer. From September to December it is peculiar that the ready made cloak and suit trade suffers severely, while any change of fashion in millinery or feathers is invariably followed by a corresponding destruction of old stock through fire. The advent of the motor car heralded the burning out of hundreds of stables, and now the influx of cheap automobiles into the market appears to approach to over production since garage outbreaks have become practically incessant. All of which is, of course, only circumstantial evidence, though it may be aptly remarked that in some countries this alone is sufficient to bring a man to the gallows.

Insurance officials argue that, in order to collect insurance on anything alleged to have been destroyed, "proof of loss" must be submitted. But for the professional firebug this matter presents no difficulty. His system of operation includes a full knowledge of whence he can obtain ample supplies of false invoices, forged affidavits and perjured testimony. In some cases, goods and furniture which have done duty in other fires are previously placed on the premises in order that all necessary proof of loss may be at hand.

This business of incendiarism is responsible to a large degree for that undesirable class of persons known technically as "public fire adjusters." It is the self-imposed duty of these functionaries immediately on the occurrence of a fire in any part of the city, to hasten to the scene and get into touch with the insured person affected by the outbreak. The keenest competition exists amongst them, and cases have been known when as many as ten were seeking the same insured party at the same time, and one of them succeeded in obtaining his client by virtually kidnapping him and carrying him away in an automobile. Ostensibly these "adjusters" play the part of philanthropists, actually they are influenced solely by motives of keen self interest. Instances have been known where such men have obtained as many as five separate contracts from an insured person immediately after a fire, each contract promising ten per cent of the insurance money to the adjuster; the assured being thus compelled on settlement to give up fully fifty per cent of his claim against the insurance companies. Although there are, no doubt, many honest agents, it is desirable to point out some of the questionable methods employed, especially in cases where arson charges are involved, thus giving direct encouragement to incendiarism. It is safe to hazard that if many incendiaries had to appear personally in the offices of insurance companies or of their accredited agents, and could not conceal themselves behind the crooked adjuster, the actual facts connected with many questionable fires would be revealed.

The most pernicious practice imaginable is that of the agent, who when he solicits business amongst known fire-bugs has a distinct understanding with them that fires are to follow the issue of policies. This incriminates these gentlemen equally with their clients and they most richly

INCIPIENT "FIRE-BUGS."

deserve a long term of imprisonment. Others again instruct policy holders how to "pad" their claims against
companies without any appreciable risk of discovery.
Hence, human nature being admittedly frail, it is not uncommon for an individual to realize that by this means he
can secure a maximum financial return for a minimum outlay.

The writer would here point out that incendiarism does
not only affect the social fabric of the community, but multi-
plies to an inconceivable degree the labours of the fire-fighting force. For generally speaking the incendiary lays his fire
in such a way that it is of an obstinate character and only
too likely to involve its surroundings. Also, it is deplor-
able to relate that women are among the most expert
in this nefarious trade; many an innocent looking curtain
and gas-jet blaze, or clothes-closet fire is the skilfully
executed work of the female incendiary. In this connection the following may be taken as illustrative of the lengths
to which women will go in their efforts to make money
by this means. During the night of August 15th, 1910,
a motorman on a trolley car passing down Third Avenue,
Brooklyn, noticed a red glare of a fire in one of the houses
on the route. With commendable curiosity he stopped and
investigated. He saw a woman, apparently sleeping, near
the doorway of a shop, with her two children beside her—
one an infant in a cradle. Being a hot night, there was
nothing particularly surprising in this. The shop door,
however, was ajar, and the motorman peeped in. A strong
smell of benzine assailed his nostrils and in his anxiety
to ascertain the cause he pushed the door further open and
stumbled upon two little bonfires blazing merrily. Promptly
arousing the apparently sleeping woman, he turned in the
alarm. Other tenants in the premises, which contained
a number of families and children, rushed down and
attempted to put out the flames. Then the "Sleeping
Beauty" of fiction became the shrew of fact, and a wicked
one to boot. "Don't do that," she screamed angrily. "You
will only spread the fire. Let the firemen put it out." Her

peculiar anxiety not to have the outbreak promptly extinguished aroused suspicion, and investigations were made.
Firemen found several wide mouthed bottles in different
parts of the shop, all containing kerosene, around their
necks being tied cords which led to a main string passing
out under the door to where this ingenious lady had been
pretending to sleep. Her explanation of the paraphernalia
was unintentionally humorous. She suggested that it must
have been the action of a "wicked" burglar. This naïve
proposition, however, did not satisfy the authorities, and
after a severe cross-examination she admitted that the fire
had been made at the instigation of a so-called adjuster.
This enterprising agent, learning that she had only thirty
cents left in the world, had glibly pointed out to her the
great advantages to be derived from a fire policy followed
by a convenient fire. He had dilated upon his success
as a professional incendiarist, remarking that in Chicago he
had engineered two uncommonly remunerative ventures.
In the first he had "made" the fire while the family, in order
to avoid suspicion, had gone to a cinematograph show,
while in the second case, in order to give some spectacular
realism to a bald piece of villainy, he had actually allowed
himself to be rescued at the crucial moment by the fire department. Acting upon this information, the police made
inquiries and quickly ran to earth the promoter of this
dastardly plot. Brought face to face with his accuser,
a dramatic scene ensued. The woman, upon it being pointed
out to her that she had endangered the lives of numerous
innocent children through the inhuman character of her
act, completely broke down and exclaimed, "I didn't want
the fire, I didn't do it! I will tell the truth to show that
I made a mistake in being influenced by a wicked man.
He is a firebug and has made many fires in Chicago."

It only remains to be said that the woman received a
well merited sentence of five years penal servitude, while
the community will be freed from the attentions of her ac-
complice for double that period. One more account of
feminine ingenuity. A lady residing in an apartment house

with her three children, had as her sole lodger an old soldier with a wooden leg. One morning she peremptorily gave him notice to leave the same day, and within twenty-four hours a regrettable, and, of course, accidental, fire gutted the flat. The insurance company concerned paid her claim without demur, the sufferer removing without delay to more commodious quarters in another part of the town. After a short sojourn there, she announced her intention of paying a visit to the seaside. The night following her departure, some children sleeping in the apartment below the one she had vacated, were awakened by hot water dripping upon them from the ceiling. Immediate investigation resulted in the discovery of a fire in the flat above, the heat of which had melted the water pipes and had thus been instrumental in arousing the inmates of the house to the peril of their position. After the fire department had suppressed the outbreak, a remarkable state of affairs was disclosed. Sideboards, cupboards and closets were found to be literally packed with ingenious "time plants," guaranteed successfully to smoulder for several hours, and then, by bursting into flame, to work their wicked will upon everything inflammable in their vicinity. Under the bed was also discovered a wooden box stuffed with papers and cotton waste soaked in oil and surmounted by the inevitable candle. In the presence of such glaring evidence the woman was obliged to cut short her holiday and return in the company of a police officer. The insurance company which had been mulcted in damages over the preceding fire suddenly bethought itself of the unusual claim of "$60 for one wooden leg," and upon making inquiries found that the possessor of this means of locomotion had never mourned its loss. Brought to trial, after a lengthy hearing, the accused was found guilty of "arson in the first degree."

The writer feels that he cannot do better than give the exact words of the judge who passed sentence upon this callous fiend. "There are certain crimes which are so revolting in their utter disregard of human life that one

wonders at the cold-blooded calculation necessary to per-
petrate them. . Such a crime is arson in the first degree,
for which crime you were indicted, and for which you
have been convicted in a lesser degree after a careful
trial—the first woman found guilty of this crime here
in twenty years. I am convinced that you were responsible
for the previous fire in your former home, and, when you
found that you were not suspected of that crime you
planned this affair, and at the same time, increased the
insurance upon your property. When the defendant is
a woman, a mother, who with fiendish indifference for
the lives of two families in her house with four little children in one and two in the other, acts as you have, such
a deed passes human understanding upon any other hypothe-
sis save that you were capable of becoming a murderess
by that midnight fire, arranged in your rooms with the
candles set in the oil soaked combustibles. You, absent to
avoid suspicion, and all for the paltry insurance money
you hoped to get. I have never seen a cooler, a more
calculating prisoner; no womanly sympathy is here, simply a fire fiend trying to secure money at any cost. Any
feeling of pity or sympathy for you at this hour I must
suspend before my stronger feeling of duty towards the
people of this community, whose lives and property have
twice been in jeopardy through your act. You are a menace to this city of homes and I therefore sentence you to
remain in prison for a term of not less than fourteen
years and not more than fourteen years and six months."

Comment upon the above is superfluous, unless it be to
say that never was sentence so richly deserved.

Because it is almost inconceivable that women should
descend to such depths, these instances of female depravity
have been given precedence in the roll of dishonour connected with incendiarism. But let it not be imagined that
the crimes of men in this direction are any less horrible
or less callous.

The story of Samuel Brant is of recent occurrence and
is one of the few instances where a firebug has been caught

red-handed. Brant openly boasted that he had worked up his profession into a high art and that no fire marshal would ever suspect him of the many charges which could be placed to his account.

With two other men he arranged to set fire to a certain flat in Brooklyn and it may have been his over-confidence which gave the clue to the ever vigilant police department. Unknown to Brant he had been under surveillance for some time and the exact hour at which the fire was to take place had been discovered. The fire marshal being in the know, arranged that several of his staff should disguise themselves as street cleaners and peddlers and loiter about in the vicinity of the premises. In a push cart, beneath a load of potatoes and other vegetables, were concealed a length of hose, some hand grenades and various fire-fighting apparatus. All these precautions were taken in order not to arouse Brant's suspicions, but just at the moment when all arrangements had been perfected, a guile-less policeman very nearly caused the ruin of the plan. He had stationed himself so near to the house in question that it was feared Brant might take alarm and make his escape. Through the medium of a woman a note was sent to the officer stating the case and asking him to leave his beat for the time being. Almost immediately after the departure of the policeman smoke was noticed to be issuing from the windows of the apartment in question, and Brant, accompanied by one of his accomplices, was seen to hurry from the house. This was the signal for the supposed street cleaners to throw aside their brooms and for the peddlers to advance nearer with their innocent looking push-cart. Rapidly they closed in on the two men, who, remarkable to say, showed fight, since the genus firebug does not as a rule suffer from a surplus of physical courage. They were quickly overcome and handed over to the police, the peddlers suddenly developing into first-class firemen, who speedily extinguished the flames. The fire had been started in a clothes closet and the flat was literally a magazine of combustible material. At his trial Brant remarked, "I am a specialist in making fires, and I can make them so that no one can catch me. The Fire Marshal is a joke. If he gets you, all you have to do is to tell him that you were away and get some one to prove it." It was proved that Brant and his associates worked a regular system. One of them would solicit business by going to the owner of a store, flat or small business concern and offer to arrange for the insurance; at the same time planning the burning of the place. His terms were somewhat exorbitant, judging at least by that operation which cost him his freedom for fifteen years. A policy had been taken out for goods supposed to be worth $800, and from this sum no less than $500 was to be deducted by way of commission, or approximately 65 per cent of the claim.

Incidentally Brant's gang was by no means unique, others are known to have operated in Chicago and Paterson, N. J., and if they have ceased from their efforts it must in no small degree be due to the active campaign waged lately against all of their kidney by Commissioner Johnson of the New York Fire Department, who can well claim to be their bitterest foe.

Undoubtedly one of the most dastardly acts in the entire history of incendiarism was the series of operations carried on during the year 1912 by a gang under the leadership of a fiend in human form known popularly as "The Torch." Their system of swindling the fire insurance companies was peculiarly atrocious, and consisted of obtaining policies on good horses, substituting for the same broken-down hacks, and then burning the latter in order to collect their claims. Fortunately, for a week prior to the night of one of their projected holocausts, the suspects had been watched and their movements had become known to the Fire Marshal. "The Torch" was regarded as a desperate character, and hence the Fire Marshal's assistants who were chosen to surround the stables involved on the night in question, were heavily armed, while some two hundred yards away two steam fire engines were stationed in readiness for immediate action. Shortly after midnight the watchers were rewarded by seeing a glare inside the stable, and a moment later "The Torch" and his son were observed making their way from the rear of the stable through a hole under the mangers. An alarm whistle was blown; three revolver shots punctuated the silence, a signal to the firemen to hurry with their apparatus, and a moment later the two desperadoes were fighting like wildcats in the hands of their captors. When an entrance into the stable had been effected, it was difficult even for men accustomed to all kinds of human rascality, to realize that what they saw was the work of men and not of devils. There were three fires burning, one just inside the doorway, a second a few feet away and another in a corner immediately behind seven helpless horses which were tethered to the mangers. The coats, tails and manes of two of these animals were saturated with gasoline. One of them was blind and the other was lame. The fire burning inside the doorway was so arranged as to block the only exit in case of possible rescue, and succeeded so well in its intention that for a considerable time it hindered and rendered most dangerous the efforts of the firemen. The actual owner of the horses confessed that he had hired "The Torch" to carry out this inhuman task, since he had been told that the latter was an expert in that line of business. With the utmost callousness this firebug admitted his share in the deal, and showed not the least emotion when told that for the next twenty years, if the world was so unfortunate as to be encumbered with his presence for that time, he would be compelled to make his home at Sing Sing prison. Though the writer knows full well the sentiments of humanitarians anent corporal punishment, he is unable to dissociate himself from a firm conviction that for crimes of this nature, perpetrated with such cold-blooded brutality, flogging is the most suitable reward. Unfortunately the number of stable fires is considerable, and the fact that approximately thirty-three per cent. of the same are listed officially as "Cause not ascertained" leads to the conclusion that they are of suspicious origin. Here surely, is sufficient food for unpleasant thought, for the hand which will apply the match to make a bonfire of a lot of dumb animals, will most assuredly not hesitate where human lives are involved. In another case, which came under the writer's notice, no less than sixty horses would have perished miserably, but for the prompt action of the fire brigade. Six separate fires, it was found, had been started in the stalls of the stable, each plant consisting of candles surrounded with kerosene-soaked straw. For perpetrators of this kind of outrage, what human punishment can be too great?

The following case is of interest as evidencing the truth that in popular phraseology "chickens invariably come home to roost."

An enterprising gentleman who had had a suspicious fire in a candy store, had been carefully kept under supervision, as it was expected that initial success would encourage future operations. One bleak March morning, a police officer was on patrol in the neighbourhood of the suspect's store when he noticed a man with a bundle of newspapers walking briskly down a side street. In a casual way he watched him and saw him throw away something which tinkled metallically as it fell on the pavement. The officer picked it up and found it to be a portion of a toy cash register made of black enameled tin. Putting it in his pocket, he resumed his patrol and a moment later came upon a motorman who had discovered a fire in the identical candy store under observation, and the alarm was turned in. The place was locked and there was a strong smell of kerosene. While waiting for the arrival of the fire apparatus, who should turn up but the same man whom the policeman had seen throw away the metal register. The store was completely gutted, and investigation clearly pointed to incendiarism. But direct proof was lacking. It was established that the owner was in serious financial difficulties, his account at the bank consisted only of six cents, and neighbors testified that his checks had been returned marked "insufficient funds." Further, shortly before the fire, he admitted that he had borrowed money. This was certainly evidence of a presumptive character, but inadequate to secure conviction. On searching the remains of the fire, however, a charred toy cash register was discovered minus the portion corresponding to that which had been picked up by the policeman. Confronted with this exhibit, the suspect first declared that he kept several of the same design for sale. Later, under cross-examination, he allowed that for fun his wife had used one and had deposited therein two dollars. The line adopted by the prosecution was that the accused had prepared his store for the fire, and that just prior to his departure he had recollected the two dollars and had broken open the register in order to secure it, carelessly throwing a portion of the same away in the street. Counsel for the defense sought to shatter this theory by producing a brand new toy register of similar design in court. Triumphantly he pointed out the following notice: "To open this bank place ten dollars in coin. It will then open automatically. If you don't deposit ten dollars in coin you will have to get an axe." Where, pleaded the counsel, was the evidence that accused had ever even possessed an axe? It was obvious that a blaze of this nature, which had not even incinerated a toy cash register could not so completely destroy a steel axe head that no trace of it could be found! And the fire department had never suggested that they had come upon any trace of such a thing!! Further, his client maintained most strongly that the policeman who identified him as the individual who had dropped the portion of the register on the morning of the fire, was in error. And in any case he defied the jury to find any cause to connect the cash box of the accused's wife with that under discussion. It had been proved that the box was unopenable without an axe,—where was the axe?

Upon this the jury retired to consider their verdict. Everything seemed in favor of the prisoner, when one of their number asked to inspect the exhibit. Within the space of three minutes he had disproved the printed statement on its exterior and had opened it with a pen-knife. That candy store keeper received a well-earned five years' imprisonment.

It would be easy to continue multiplying instance upon instance and story upon story to show that the existence of the working incendiary is no figment of the writer's imagination, but rather a fact with which municipalities, fire departments and insurance companies have got to grapple. It accounts in part for the remarkable discrepancies between fire losses in American cities and those in European communities. During 1910 London had 3,941 fires, Paris 2,030, Berlin 2,068, and New York 14,405. For every one hundred thousand inhabitants Berlin has 97 fires, London 81, St. Petersburg 75, Paris 74, Vienna 59, and New York 300. The fire loss per head of population in the United States generally is nearly five times greater than that of any foreign country. In New York, during 1911, the per capita loss was $2.45, while the average for European cities was about $0.50, sinking as low as twelve cents in two towns so differently situated as Southampton and Dresden. After making every allowance for climatic differences, structural defects and the use of inflammable building materials, it is difficult to escape the conclusion that the firebug has a lot for which to answer. Broadly speaking, it is not exaggeration to estimate twenty-five per cent. of New York fires certainly, as of incendiary origin. The insurance risks carried by the one hundred and seventy-five companies in New York total the gigantic figure of forty billion dollars, spread throughout the country. Hence, it goes without saying that the influence exerted by these corporations, financial and otherwise, is stupendous and may indirectly control the welfare of the community. There are not wanting those who maintain that insurance companies, within a certain degree, welcome fires as bespeaking business. It is reported that the manager of a Scottish insurance company in a speech at Edinburgh, said, "Were there no fires there would be no insurance business. And on the other hand, the greater the fire damage, the greater the turnover, out of which insurance companies make profits." Now this is only the report of a speech, and quite probably has been transmitted incorrectly, for it most certainly is at variance with the opinions of the insurance officials with whom the writer has come in contact. Rather is the question one affecting the nation as a whole. The search after all classes of business is so keen nowadays, the turnover so tremendous and the demands of the share-holders for large profits so exacting, that directors and others responsible must be pardoned if in their anxiety to do the best for those dependent upon them they accept risks which cooler calculation and difference of environment would show to be preposterous. It seems absurd to discuss an evil and then not to suggest the remedy. But incendiarism, though actively affecting the routine of fire departments and causing fire chiefs endless worry and anxiety, belongs properly to a sphere outside the purview of the scientific fire-fighter. It is an excrescence on the social fabric which needs removal by those specially equipped for the task. And undoubtedly, those referred to are the insurance companies. The means and methods to be employed must be left to them, for it would be as futile for the writer to tender suggestions on such a highly complicated problem, as it would be absurd for underwriters to give advice to him regarding the best way to fight a fire in a warehouse filled with explosives. But it is satisfactory to be able to state that already signs are not wanting of a general awakening of interest in the subject amongst all classes affected, professional and otherwise. That is to say the insurance companies are on the move and it is no longer so easy to effect policies on worthless goods, while the individual of doubtful financial stability and dubious reputation is likely to experience considerable difficulty in persuading even the most reckless of agents to consider seriously his application. Towards this happy consummation, no one has worked with more energy and good will than Commissioner Johnson of the New York Fire Department, to whose publication on the subject the writer is indebted for many illuminating facts used in this chapter. It will at least be conceded by all concerned that the introduction of legislation to assist the insurance companies in their laudable efforts by "making the punishment fit the crime" and thoroughly frightening the firebug by the penalties awaiting him, would be a distinct step in the right direction.

CHAPTER IX

·

THE advent of the motor car has not proved an unmixed blessing to the fire-fighter, and it is no exaggeration to say that the general adoption of motor traction has enormously increased the fire-risk. In the first place, gasoline, the most usually employed of motor oils is an extremely dangerous substance to handle, though that familiarity which breeds contempt has robbed it of its sinister significance, while ignorance of an almost culpable nature has rendered its handling additionally and unnecessarily perilous.

The first essential for motor owner, chauffeur or garage proprietor is that he should understand something of the chemical qualities of gasoline, in which term may be included all other spirits of a kindred nature such as petrol, naphthaline, etc. This does not mean that they must study the subject with the microscopic care of the professional chemist, but it does presuppose that any individual gifted with common sense prefers to know the characteristics of the most important adjunct of the machine he essays to own, drive or house. Gasoline in its primitive state is one of the component factors forming crude petroleum. By distillation it is purified to a greater or lesser extent, automobiles, as a rule, demanding the most refined spirit available. It is possessed of no flash point, that is to say, if placed in an open vessel it will vaporize at any ordinary temperature, in fact even with the thermometer at zero. The weight of its gas is three and a half times greater than air, which forms an inherent hazard, since, unlike ordinary lighting and acetyline gases, which rise and are carried off by a breeze or through any opening which causes a draught, it falls to the floor and will lie and collect unless disturbed. Should the disturbance take the form of a lighted match or candle' a tremendous explosion results and fire follows. But the point is, that there is nothing to show that it is collecting in any particular place; it remains dormant and unobserved like a snake in the grass and is every inch as dangerous in its effects. Further, unmixed with air, this vapor is comparatively harmless; its virility depends upon its admixture with the ethereal gases, when one pint of gasoline is sufficient to make two hundred feet of highly explosive mixture.

In the liquid state, gasoline is innocuous, that is to say so long as it remains an absolute liquid it can neither ignite, burn or explode. Similarly, pure gasoline vapor will neither ignite nor burn, but requires the assistance of the air, and it is precisely for this reason that the carburetor plays such an important part in the mechanism of the motor engine. Its highest point of explosive violence is reached when roughly one part of vapor mixes with eight parts of air, and decreases in combustibility with an increase of either air or gasoline. Another peculiar property in gasoline to be noted is, that even when vaporized and mixed with air, it has a definite temperature of ignition, just as wood or any other combustible material.

Hence, it will be seen, that this spirit is often more dangerous than even gunpowder or dynamite, inasmuch as the latter will stay where they are placed while the former may vaporize, and, creeping subtly along a floor or passage, may be ignited a hundred feet or so distant from its source. The resultant flash will travel back through the gas strata, thus causing an explosion or fire at the point of its inception. With such ever-present risks attendant upon its use it might be imagined that every possible precaution would be adopted by those handling it. And yet exactly the reverse is the case.

Of all careless persons, chauffeurs and employees of

BEFORE THE ALARM IS TURNED IN.

garages may justly claim preëminence. In spite of printed regulations and orders prominently displayed they will smoke with the utmost insouciance at every possible opportunity, absolutely heedless of the fact that they would be just as well advised to smoke in a powder mill! And if the employees are bad then the owners are not much better. Unless compelled by municipal ordinances, they are sublimely indifferent to effective fire protection in their garages, and with the slightest encouragement will press into their service any building, however unsuited to the purpose, either by structure or convenience. An empty stable, a disused church, a ramshackle warehouse built of wood, anything does so long as there is sufficient floor space and there is any method by which the law can be contravened with impunity. These are some of the difficulties which the modern fire-fighter must be prepared to encounter and by some means overcome. Needless to say, drastic laws have been introduced for the proper storage of gasoline in garages, though in this direction a very curious anomaly may be noted. Thus, while the gasoline in the main tank is assiduously protected, no attention is given to the spirit in the tanks of the automobiles themselves, often amounting to thirty or forty gallons per tank and located haphazard throughout the entire building. It is obvious that, if a fire starts, such an arrangement is only too likely to lead to disaster, and that the care displayed over the main gasoline tank is not unlike locking the windows against burglars and leaving the door wide open. Broadly speaking, gasoline should be stored in a well-made tank, underground and beneath the floor of the garage, and in this connection it will be apropos to give some excerpts from the regulations governing garages and the storage of gasoline in New York city.

The following six sections explain succinctly where garages should under no circumstances be situated. (A) No garage must be within fifty feet of the nearest wall of a building occupied as a school, theatre or other place of public amusement and assembly. (B) It must not be situated in any building occupied as a tenement house or hotel. This is by no means uncommon in some parts of Europe, though any one conversant with the peril he is running would preferably sleep above a fireworks factory. (C) Garages may not be located in buildings not constructed of fire-resisting material throughout. (D) They may not be situated in places where paints, varnishes or lacquers are either manufactured, stored or kept for sale. (E) Or where drygoods or other highly inflammable materials are manufactured or kept for sale. (F) Or where rosin, turpentine, hemp, cotton, gun cotton, smokeless powder, blasting powder, or any other explosives are stored or kept for sale.

Such regulations may sound absurd to the average citizen. Who on earth would want to have a garage in a place where explosives are stored? it may be asked; and though this may be extreme it is a fact that most of the regulations framed for fire protection are fashioned to guard against the proved thoughtlessness of the individual. The writer is reminded of a genial character he encountered once in his travels in a certain West African port. The gentleman in question casually knocked his pipe ashes out against the rim of an open keg of blasting powder. The remonstrances of his mates, which were of a physical nature, elicited from him the excuse, "Well, I've often done it before and nothing has ever happened." It was quite useless to argue the point; that he would have been blown to Jericho, or somewhere else, but for the mercy of providence weighed with him not a whit. It is persons of this type who make "nursery" legislation necessary, and their name in the motor world is legion.

The following sections explain themselves and serve to illustrate how gasoline should be stored, having due regard to safety. (A) "Each storage tank shall be constructed of steel at least $\frac{1}{4}$ of an inch thick; shall have a capacity of not more than 275 gallons, and shall, under test, stand a hydrostatic pressure of at least 100 pounds to the square inch. (B) Each storage tank shall be coated on the outside with tar or other rust-resisting material, shall rest upon a solid foundation and shall be embedded in and surrounded by at least twelve inches of Portland cement concrete, composed of two parts of cement, three parts of sand and five parts of stone. (C) Each storage tank installed in a garage shall be so set that the top, or highest point thereof, shall be at least two feet below the level of the lowest cellar floor of any building within a radius of ten feet from the tank." Garages constructed along these lines are unlikely readily to catch alight, and the financial outlay rendered necessary by such structural additions is as nothing to the increased security obtained.

The following rules should also be rigidly observed and are applicable to garages attached to private houses, which, be it said, are often carelessly looked after since both master and man are only too prone to be lax, especially when outside the sphere of city regulations. Incidentally, however, this is precisely one of the occasions demanding the maximum of precaution. "All oils spilled on the floors of a garage should be removed at once by sponging or swabbing, and should be poured into the drain leading to the oil separator which is installed so as to be connected to the house drain, and so arranged as to separate all oils from the drainage of the garage.

"No system of artificial lighting other than incandescent electric lights should be installed in any garage unless of a type for which a certificate of approval has been issued by the fire commissioner." Of course, in the country, there may be some difficulty over this provision, but common sense applied to the problem will certainly limit the fire risk. It also goes without say that no stoves or any appliance likely to produce an exposed spark should be installed in a garage, unless placed in a room separated from it by fireproof floors and walls. As regards the carelessness of the individual, the following excerpt taken from a speech made at the annual meeting of the National Board of Fire Underwriters, needs no comment.

"I confess it is astonishing to find that the fire waste

is not diminished by the better character of buildings we are getting. We are getting better buildings than we ever did before, but the losses keep up, and this is because fires cost more today than they ever did before. And there are new hazards. We are using higher explosives; we are using higher potentials in electrical practice; we are using more gases, like gasoline. Ten years ago the gasoline engine was a clumsy device; there were but few. The development of the gasoline engine has brought a widespread field for it. The farmer uses it for cutting his feed and grain; the merchant uses it; the manufacturer uses it. The automobile has scattered gasoline all over the country. To my desk there come reports of thousands of fires every year from gasoline—cleaning with gasoline, garages stored with gasoline, and the cheerful idiot who smokes cigarettes in the garages and throws matches about. Useless, unnecessary fires must be checked. If we can place individual responsibility; if we can change the attitude of the people toward the man who has a fire so that they can see that he is not an object of sympathy but a man who has offended against the common welfare, unless he can prove that he was in no way responsible for that fire, then we will approach the time when we can diminish those hazards. That point of view must be emphasized, and when every man who has a fire will have to step up before the Fire Marshal's investigation and is exhibited to his fellows as an offender against the common good, as a picker of the pockets of the rest of us, I believe we will correct these habits of carelessness."

The writer cordially endorses the above and, as regards fire control in garages, is inclined to add that for the lax in this respect, no condemnation can be too severe.

From the latest report of the New York Board of Fire Underwriters it appears that of 206 recent fires 33 per cent. were due to the use of gasoline for cleaning cars and 43 per cent. were due to back fire into the carburetors of automobiles. Amongst the others were 5 from filling tanks of automobiles with lamps burning, 3 from smoking,

SMOKY FIRE, NEW YORK.

4 from gasoline leaks in contact with a hot exhaust pipe, 5 from defective electric equipment on cars and 1 from spontaneous combustion. These figures point to the fact that the promiscuous use of gasoline in many garages for cleaning purposes, taken in conjunction with the number of fires attributed to this cause, is one of the most serious hazards with which to contend. Although the investigations indicate that 33 per cent. of all fires of known cause were due to this practice, the actual number is probably even greater as there is reason to believe that an appreciable number of fires reported as caused by back fire into carburetors are due directly or indirectly to cleaning parts of the car with gasoline. In a number of the best managed garages the prohibition of the use of gasoline for cleaning purposes is strictly enforced, and the use of oils no more volatile than kerosene is insisted upon. In other cases even kerosene is prohibited for such purposes and use of caustic soda and water or a similar solution is required. One golden rule for all garages, public or private, is that a number of buckets filled with sand should be kept in readiness for any emergency, while in the way of hand extinguishers those containing carbonate of chloride are amongst the most effectual.

Another fruitful source of danger, as far as the use of gasoline is concerned, is its employment in dry-cleaning and sponging establishments. In fact, it is an interesting commentary upon the philosophy of life that those elements which are of the greatest general use to society are nearly always fraught with an irreducible minimum of risk, if applied without caution. The cleansing properties of gasoline are beyond estimate; upon this being discovered, fools literally stepped in where angels feared to tread, with the result that several lives were lost in consequence of hairdressers using this spirit as a shampoo, while it was not unusual for employees in dry-cleaning establishments to wander around gas lighted rooms with trays full of the liquid.

Things have altered since then. The former operation

has been forbidden, and the latter is now hedged in with such restrictions that safety is, to a considerable extent, guaranteed. Usually the method employed consists of revolving drums, each containing thirty or more gallons of gasoline, which, being in a constant state of disturbance, has a tendency to throw off heavy fumes, hence the drums must be kept closed. When the garments are removed and placed in the rotary driers, or centrifugals, more fumes are given off; and, finally, the function of the drying room is to enable the clothes to throw off such gasoline as still remains in them, so that this room is especially thick with vapor. In addition, a number of open vessels containing from five to fifty gallons of spirit will be found scattered about the place, their *raison d'être* being to facilitate the cleaning of gloves, laces and other light and filmy fabrics. The hazard in places of this description is too apparent to require much elaboration, and it need only be said that the system of storing the main supply of gasoline should be the same as in garages, namely underground. In this connection it is of interest to note that never, in the experience of the writer, has any fire started from an underground storage system, and in no case has fire been increased because of such a system. In fact, there is no case on record where the gasoline in a buried tank has been affected by a fire. This proves conclusively that there is no danger in its storage when properly arranged, but only in its handling. Thus, the latter should be expedited in every possible way, and so arranged that the gasoline is not exposed to the air, and the ventilation of garages and dry-cleaning plants should be so effected that no gases can accumulate on the floors.

Hence, the safe and sane handling of gasoline is no longer a question of insurmountable or insuperable difficulty. Inasmuch as the automobile has come to stay, inasmuch as motor traction will be increasingly applied in the near future for all classes of transportation, and inasmuch as the same familiarity, akin to the affection formerly shown to the horse, will now be extended to the motor car, though the affection for the former must not be allowed to develop into contempt for the latter, then it behooves the layman to understand something of the tool with which he will be called upon to deal. Gasoline has been termed "man's unseen enemy," but, like many other potential adversaries, careful handling may transform it into a useful servant and a trusty friend.

In conclusion, in order to emphasize the point once again, that point which is so regularly neglected and which is such a fruitful source of danger to the community at large, the words of the New York Fire Ordinance may be quoted *in extenso;* they apply to all places in which gasoline is either used or stored. "It shall be unlawful for any person to smoke or to carry a lighted cigar, cigarette or pipe, into any room or compartment in which volatile or inflammable oil is stored or used; and a notice bearing in large letters the words 'SMOKING FORBIDDEN,' together with an excerpt of the rules governing the subject in smaller letters shall be displayed in one or more conspicuous places on each floor where volatile inflammable oil is stored or used. Those breaking the regulation hereon displayed are guilty of a misdemeanor."

CHAPTER X

PART I

GREAT conflagrations are plentifully recorded during Roman times and, as has been shown, all that the science of the period coupled with most commendable forethought could accomplish, was done to stave off the peril. None the less, however, the magnificent "Basilica Julia," a building devoted to law courts, completed by Augustus in B.C.44 after plans designed by Julius Cæsar, was entirely gutted and remains to this day a relic of architectural antiquity and a perpetual reminder that fire risks ever were, and probably ever will be, amongst the perils of existence. Again in 64 A.D. Rome was devastated by an outbreak which lasted three days and burned out most of the residential portion of the city. It has been popularly attributed to that peculiarly eccentric emperor Nero, but in justice to that despot it must be added that the evidence of his being a "firebug" on a gigantic scale is slight. Then occurred a lapse of centuries, during which, no doubt, bad fires took place, but they were not of a sufficiently startling character to leave any permanent mark upon history till the partial destruction of London in 1666. The details of this conflagration are so well known that it seems almost unnecessary to dwell upon it, but the following description drawn from a diary of that gossipy old chronicler Samuel Pepys, appears worthy of quotation, since he was an eye-witness, and the style in which he writes is so quaint:

"Sept. 2nd. Lord's Day. Some of our maids sitting up late last night to get things ready against our feast to-day, Jane called us up about three in the morning to tell us of a great fire they saw in the city. So I rose and slipped on my night gown and went to the window; but being unused to such fires as followed, I thought it far enough off; and so to bed again and to sleep. . . Bye and bye Jane comes and tells me that she hears that above three hundred houses have been burned down tonight by the fire we saw and that it is now burning down all Fish Street by London Bridge. So I made myself ready presently and walked to the Tower; and there got up upon one of the high places, Sir J. Robinson's little son going up with me; and there I did see the houses at that end of the bridge all on fire and an infinite great fire on this and the other side, the end of the bridge; which among other people did trouble me for poor little Michell and our Sarah on the bridge. So down with my heart full of trouble to the Lieutenant of the Tower, who tells me that it begun this morning in the King's baker's house in Pudding Lane, and that it hath burnt down St. Magnus's Church and the most part of Fish Street already. So I go down to the waterside and there got a boat, and through bridge, and there saw a lamentable fire. Poor Michell's house as far as the Old Swan, already burned that way, and the fire running further, that in a very little time it got as far as the Steele-yard, while I was there. Everybody endeavouring to remove their goods, and flinging into the river, or bringing them into lighters that lay off; poor people staying in their houses as long as till the very fire touched them, and then running into boats or clambering from one pair of stairs by the waterside to another. And among other things, the poor pigeons, I perceive, were loth to leave their houses, but hovered about the windows and balconies till they burned their wings and fell down. Having staid, and in an hour's time seen the fire rage every way; and nobody to my sight endeavouring to· quench it, but to remove their goods and leave all to the fire; and having seen it get as far as the Steele-yard, and the wind mighty high and driving it into the city; and everything after so long a drought proving combustible, even the very stones of the churches; and among other things the poor steeple by which pretty Mrs. ———— lives, and whereof my old schoolfellow Elborough is parson, taken fire in the very top and there burned till it fell down; I go to Whitehall with a gentleman with me, who desired to go off from the Tower to see the fire in my boat; and there up to the King's closet in the Chapel where people come about me, and I did give them an account, dismayed them all, and word was carried into the King. So I was called for and did tell the King and the Duke of York what I saw; and that unless his Majesty did command houses to be pulled down, nothing could stop the fire. They seemed much troubled, and the King commanded me to go to my Lord Mayor for him, and command him to spare no houses, but to pull down before the fire every way. The Duke of York bid me tell him, that if he would have any more soldiers, he shall; and so did my Lord Arlington after, as a great secret. Here meeting with Captain Cocke, I in his coach, which he lent me, and Creed with me to St. Paul's; and there walked along Watling Street as well as I could, every creature coming away loaded with goods to save, and here and there sick people carried away in beds. Extraordinary good goods carried in carts and on backs. At last met my Lord Mayor in Canning Street, like a man spent, with a handkercher about his neck. To the King's message he cried, like a fainting woman, "Lord, what can I do? I am spent: people will not obey me. I have been pulling down houses; but the fire overtakes us faster than we can do it." That he needed no more soldiers; and that, for himself, he must go and refresh himself, having been up all night. So he left me and I him and walked home; seeing people all almost distracted, and no manner of means used to quench the fire. The houses, too, so very thick thereabouts, and full of matter for burning as pitch and tar in Thames Street; and warehouses of oyle and wines and brandy and other things. Here I saw Mr. Isaac Houblon, the handsome man, prettily dressed and dirty at his door at Dowgate, receiving some of his brother's things, whose houses were on fire; and, as he says, have been removed twice already; and he doubts, as it soon proved, that they must be removed from his house also, in a little time, which was a sad consideration. And to see the churches all filling with goods by people who themselves should have been quietly there at the time. By this time it was about twelve o'clock and so home and there find my guests.

"So near the fire as we could for smoke; and all over the Thames, with one's faces in the wind, you were almost burned with a shower of firedrops. This is very true; so as houses were burned by these drops and flakes of fire, three or four, nay five or six houses, one after the other. When we could endure no more upon the water, we to a little alehouse on the Bankside, over against the Three Cranes, and there staid till it was dark almost, and saw the fire grow; and as it grew darker, appeared more and more; and in corners and upon steeples, and between churches and houses, as far as we could see up the hill of the city, in a most horrid malicious bloody flame, not like the fine flame of an ordinary fire."

The chronicler at this point is forced to leave his own home and finds shelter with one, Sir W. Rider. This occupied him during the 3rd of September and he continues on the 4th.

"Sir W. Pen and I to the Tower Street, and there met the fire burning, three or four doors beyond Mr. Howell's, whose goods poor man, his trayes and dishes, shovells, etc., were flung all along Tower Street in the kennels, and people working therewith from one end to the other; the fire coming on in that narrow street with incredible fury. . . . And in the evening Sir W. Pen and I did dig another (pit) and put our wine in it, and I my parmazan cheese, as well as my wine and some other things. . . . I after supper walked in the dark down to Tower Street, and there saw it all on fire, at the Trinity House on that side and the Dolphin Tavern on this side, which was very near us, and the whole heaven on fire. Now begins the practice of blowing up of houses in Tower Street, those next the Tower, which at first did frighten people more than anything; but it stopped the fire where it was done, it bringing down the houses to the ground in the same places they stood, and then it was easy to quench what little fire was in it, though it kindled nothing almost. . . . 5th. About two in the morning my wife calls me up, and tells me of new cryes of fire, it being come to Barking Church, which is the bottom of our lane. I up, and finding it is so, resolved presently to take her away, and did, and took my gold, which was about 2,350 pounds, W. Hewer and Jane down by Proundy's boat to Woolwich; but Lord, what a sad sight it was by moonlight, to see the whole city almost on fire, that you might see it as plain at Woolwich, as if you were by it. . . . But to the fire, and there find greater hopes than I expected; for my confidence of finding our office on fire was such, that I durst not ask anybody how it was with us, till I come and saw it was not burned. But, going to the fire, I find, by the blowing up of houses, and the great help given by the workmen out of the King's yards, sent up by Sir W. Pen, there is a good stop given to it, as well at Marke Lane End as at ours; it having only burned the dyall of Barking Church, and part of the porch, and was there quenched. I up to the top of Barking steeple, and there saw the saddest sight of desolation that I ever saw; everywhere great fires, oyle cellars, and brimstone and other things burning. I became afraid to stay there long, and therefore down again as fast as I could, the fire being spread as far as I could see it; and to Sir W. Pen's, and there eat a piece of cold meat, having eaten nothing since Sunday but the remains of Sunday's dinner. Here I met with Mr. Young and Whistler, and having removed all my things, and received good hopes that the fire at our end is stopped, then I walk into the town and find Fenchurch Street, Gracious Street, and Lumbard Street all in dust. The Exchange a sad sight, nothing standing there, of all the statues or pillars, but Sir Thomas Gresham's picture in the corner. Into Moore-fields, our feet ready to burn, walking through the town among the hot coals, and find that full of people, and poor wretches carrying their goods there, and everybody keeping his goods together by themselves; and a great blessing it is to them that it is fair weather for them to keep abroad night and day; drunk there, and paid two-pence for a penny loaf. Thence homeward, having passed through Cheapside, and Newgate market all burned; and seen Anthony Joyce's house in fire; and took up, which I keep by me, a piece of glass of the Mercer's chapel in the street, where much more was, so melted and buckled with the heat of the firelike parchment. I did also see a poor cat taken out of a hole in a chimney, joyning to the wall of the Exchange, with the hair all burned off the body and yet alive. . . . 6th. Up about five o'clock, and met Mr. Gauden at the gate of the office, I intending to go out, as I used every now and then, today to see how the fire is, to call our men to Bishopsgate, where no fire had yet been near, and there is now one broke out; which did give great grounds to people and to me, too, to think that there is some kind of plot in this, on which many by this time have been taken, and it hath been dangerous for any stranger to walk in the streets, but I went with the men, and we did put it out in a little time; so that that was well again. It was pretty to see how hard the women did work in the cannells, sweeping of water; but then they would scold for drink, and be as drunk as devils. I saw good butts of sugar broke open in the street, and people give and take handfuls out, and put into beer, and drink it. And now all being pretty well, I took boat, and over to Southwarke, and took boat on the other side of the bridge, and so to Westminster, thinking to shift myself, being all in dirt from top to bottom; but could not there find any place to buy a shirt or a pair of gloves, Westminster Hall being full of people's goods, those in Westminster having removed all their goods, and the Exchequer money put into vessels to carry to Nonsuch; but to the Swan and there was trimmed: and then to White Hall, but saw nobody; and so home. A sad sight to see how the river looks; no houses nor church near it, to the Temple where it stopped. And home, did go with Sir W. Batten, and our neighbour,

Knightley, who with one more was the only man of any
fashion left in the neighbourhood thereabouts, they all removing their goods, and leaving their houses to the mercy
of the fire. . . . Thence down to Deptford, and there with
great satisfaction landed all my goods at Sir G. Carteret's
safe, and nothing missed I could see or hear.   But
strange it is to see Clothworker's Hall on fire these three
days and nights in one body of flame, it being the cellar
full of oyle. 7th. Up by five o'clock; and, blessed be God,
find all well; and by water to Pane's Wharfe. Walked
hence, and saw all the town burned, and a miserable sight
of Paul's Church, with all the roofs fallen and the body of
the quire fallen into St. Faythe's; Paul's School also, Lud-
gate, and Fleet Street. My father's house, and the church,
and a good part of the Temple alike. . . . This day our
Merchants first met at Gresham College, which, by proc-
lamation, is to be their Exchange. Strange to hear what
is bid for houses all up and down here; a friend of Sir W
Ryder's having a hundred and fifty pounds for what he
used to let for forty pounds per annum. Much dispute
where the Custom House shall be; thereby the growth of
the city again to be foreseen. . . . People all over the
world do cry out of the simplicity of my Lord Mayor in
generall; and more particularly in this business of the fire,
laying it all upon him. . . . Much good discourse; among
others, of the low spirits of some rich men of the city in
sparing any encouragement to the poor people that wrought
for the saving of their houses. Among others, Alderman
Starling, a very rich man, without children, the fire at next
door to him in our lane, after our men had saved his house,
did give two shillings and sixpence among thirty of them,
and did quarrel with some that would remove the rubbish
out of the way of the fire, saying that they had come to
steal. . . . 15th. Captain Cocke says he hath computed that
the rents of the houses lost this fire in the city comes to
six hundred thousand pounds per annum. . . . 17th. By
water, seeing the city all the way, a sad sight indeed, much
fire being still in."

So much for the story of the fire of London as told by so inquisitive and garrulous an eyewitness as Samuel Pepys. He could have had no idea that two and a half centuries later, all that he remarked as passing strange would be repeated in another continent and amongst buildings higher than the then summit of "Paul's Church."

And yet it is curious to note how identical in many respects are the great conflagrations of today. The general rush for safety with never a moment's consideration as to whether after all there may not be some advantage in the defence of the home by the individual, the starting of subsidiary fires by burning embers, the use of explosives as a means of stopping a conflagration, often only to increase the damage, the frantic appeal to the Mayor to do something and the failure of that individual often to rise to the occasion and finally, of course, the finding of a suitable scapegoat upon whom to heap blame. It also proves the lamentable condition to which the science of fire prevention had sunk, when the most important and the most wealthy city of the period, not only possessed no organized plan of fire resistance, but was content to let it burn for aught its inhabitants cared, so long as their individual property was saved. The lesson, however, was not forgotten, and undoubtedly the modern fire department owes its renaissance from Roman times to this disaster, which once and for all taught the good burgesses of London and elsewhere that fire was an enemy as crafty and as dangerous as any on land or sea.

Amongst great conflagrations, that of the city of Baltimore, which occurred on Sunday, February 7, 1904, and continued over the greater part of the following day, attains special prominence from the fact that in spite of the stupendous damage done to property no lives were lost. The burnt area covered 140 acres and comprised 80 city blocks in the business section, while no less than 27 great buildings of fire resistive construction were completely gutted and, in some cases, collapsed. It may here be stated that at no time was there any shortage of water, which,

of course, is one of the most general causes for the spread
of a fire. At 10.48 on that Sunday morning, the automatic
alarm registered a call from the basement of the Hurst
Building, a wholesale drygoods house with a varied stock,
including a large supply of celluloid novelties. Its location
was the southeast corner of Liberty and German Streets
and within 48 seconds of the alarm, an engine company
and a hook and ladder company under command of the
District Chief were upon the scene. At that time no fire
was visible on the first floor, and neither smoke nor heat
was apparent. Presumably this led to an underestimation
of the seriousness of the outbreak, as the firemen promptly
proceeded to attack only with a single line of chemical hose,
passed from the German Street side of the building into
the basement. The small blaze discovered there, and probably caused by a smouldering pile of rubbish, suddenly
burst into flame, which, with incredible rapidity, ran up the
elevator shaft, driving the firemen from their positions.
About seven minutes later a violent explosion occurred,
blowing out the windows in the building and shattering all
the glass in the immediate neighbourhood. It was then
seen that the entire house was alight from top to bottom
and the flames shooting out through the windows greedily
licked the walls of the buildings opposite, which, in their
turn, took fire.

Being Sunday, a large proportion of the population were at church, when the muffled boom of the
explosion was heard above the solemn strains of sacred
music. What it portended none could tell, but in the
twinkling of an eye ministers and their congregations had
left their devotions and hurried into the street. As though
in answer to their worst fears, another dull rumble of
threatening significance was borne across the morning
breeze. Later this was ascertained to have been caused
by the explosion of a large quantity of blasting powder,
which, by blowing out more windows, expedited the onrush
of the flames.

Residents in the hilly portions of the city, gazing

fearfully in the direction of the sound, could see huge volumes of fleecy smoke rising sullenly from the business quarter, and then at last the realization was brought home upon them that they were face to face with a great conflagration.

Amongst the first to reach the outbreak were scores of business men intent upon saving their books and records, and who eagerly enlisted the services of boys, loafers, longshoremen, in fact, any person willing to aid in the all-important task. The Express Companies likewise responded with all speed to the sudden demands made upon them and sent emergency calls for all their employees to requisition hand-carts and wagons.

Meantime the outbreak had increased alarmingly, and had obviously grown beyond the control of the fire department. A district alarm had almost at once been sent in, and the Departmental Chief, hurrying to the scene of operations, had quickly realized that the flames, fanned by an increasing wind and spreading in two directions, would need a greater force to deal with them than he had at his disposal. Also bad luck seemed to dog their most desperate efforts. An attempt to save a valuable piece of apparatus cost precious time and was unsuccessful, while Chief Horton himself was unfortunate enough soon after his arrival to be incapacitated for duty by a severe electric shock from a fallen cable. It is impossible to estimate the moral effect of such an occurrence, for even as on a battlefield soldiers look for encouragement and stimulus to their commander, even more so do the rank and file of a fire-fighting force depend on the example and propinquity of their Chief.

As soon as it became clear that the conflagration was assuming colossal proportions, urgent messages were sent to surrounding towns, such as Washington, Chester, York and Philadelphia, for their assistance and ultimately even to New York, which responded to the call with promptitude. Owing to the congestion of apparatus, however, the crowds of spectators and the general confusion, many of the out-of-town engines could not be utilized to the best of advantage, while difference in hose couplings obliged numbers to obtain their own water supply direct from the harbour, thus preventing their presence where most urgent. The fire generally took a westerly direction, and the buildings in the path of the flames failed to offer any resistance, owing to their "fire-walls" being parallel to the onset.

In the town itself the conditions were lamentable. At the City Hospital, the Sisters of Mercy with smiling faces and sinking hearts endeavoured to keep all news of the fire from their charges, while the staff physicians stationed themselves on the roof in order to extinguish the burning embers which rained upon them. Finally it was deemed necessary to transport the sufferers to a place of safety in the upper town, a task carried out with the greatest tenderness and skill. Needless to add, all medical men in the town had offered their services, and though happily these were required in only a few instances, the knowledge of the fact went a long way towards reassuring the timid. From 5 o'clock in the afternoon till midnight, the fire made its greatest headway, the wind during this period having increased from 14 miles an hour in a westerly direction to twenty-five miles, after which it veered to northwest, and remained in that quarter with decreasing velocity till the finish. The spread of the conflagration in the direct path of the wind was practically unchecked by the operations of the fire-fighters, by the doubtful expedient of dynamiting both burning and unburned buildings, by the streets or by the so-called fire-proof buildings.

Minor explosions, however, did much to hamper the efficacy of the department, 152 whiskey barrels, for instance, caught fire and burst, flooding the street with burning spirit and causing indirectly the destruction of three pieces of apparatus. It may be here mentioned that valuable assistance was rendered by volunteers, numbering some two hundred, who extinguished a large number of subsidiary fires started by burning brands.

In quarters not in the direct path of the wind some successes were registered and·served to cheer the drooping spirits of the fighters. On the west side of Liberty Street, and even in the vicinity of the Hurst Building, a strong force concentrated to windward succeeded in saving a large shirt factory, keeping the temperature down to a point where the automatic sprinklers were not called into play. Subsequently that system certainly proved its value. The drygoods store of O'Neil & Co., the entire interior of which was provided with that apparatus, was threatened with destruction, the roof boards being ignited owing to their tin sheathing becoming red hot. Fifteen sprinkler heads opened and prevented that fire from spreading.

Another notable instance of successful defense was that made by a third wholesale drygoods house, the Lloyd Jackson Co., situated at the southeast corner of Liberty and Lombard Streets. Owners and employees put up a stiff fight, kept the roofs wet by hose streams from their private fire pump, and hung blankets soaked with water over the cornices. At the same time water was pumped into the sprinkler supply tank above the roof until it overflowed, when, by plugging up the roof drain pipes, the water was forced to run over the cornice and thus formed a "water curtain" down the north front of the building. A large amount of glass was broken, but there was practically no damage to the interior.

Perhaps the most dramatic scenes were enacted in the neighbourhood of the docks, where, as already stated, the out-of-town departments were able to find full scope for their services. No one lacked for employment. In the river tugs of all sizes dashed in and out amongst the shipping, towing to safety great vessels and their valuable cargoes, whose charterers or agents had visions of their entire destruction. Rescue had come none too soon, for the decks of many had grown so hot that it was agony for the sailors to tread their scorched surfaces, while the paint on funnels and sides blistered and peeled off in flakes.

A North German Lloyd cargo steamer making its way slowly up the bay, was confronted with the spectacle of what would have awaited it, had it docked a few hours earlier, and anchored hurriedly at a safe distance. One busy tug was the means of rescuing the President of the C. A. Gambril Co., whose offices were behind the fruit wharves. Absorbed in saving his books, he had not observed that his way to the street was cut off by the advancing flames until he reached the door. His only hope now lay in the docks, which were already in a precarious state, and clutching his treasures under his arms, he ran at full speed along the wharf's edge searching with anxious eyes for a boat and even meditating the final arbitrament of the water below him. Fortunately his plight was noticed, and he was dragged on to the tug none the worse for his adventure.

And now occurred the first notable victory of men against fire in this portion of the city. Had the flames succeeded in involving Denmead's malt house, not all the fire departments in America could have stemmed the tidal wave of destruction which would have ensued, and it is to the credit of the fire-boat "Cataract" that this catastrophe was averted. Aided by companies on land, she fought the oncoming conflagration with grim determination until the safety of the malt house was assured. By this time, thirty-six companies, a police boat and two tugs had concentrated all their force in the vicinity of Jones Falls, a little dirty, bad smelling stream, which had never served a useful purpose, and which the municipality had proposed filling in owing to its insanitary condition. There city stood by city, Wilmington by Chester, York by Washington, Baltimore by Philadelphia and New York, which had arrived late upon the scene but was doing yeoman's service. Five firemen on the roof of one building had a narrow escape. Working like demons to save the adjoining houses they heard shouts of warning from their comrades in the street, and to their dismay saw the flames beneath them. A tall telegraph pole, which fortunately rose to the height of the roof, on which they stood, was the only means of escape from the furnace, which they could hear roaring below them. Reaching a tin gutter, which afforded them some hold, they, one by one, clutched the pole and slid to the ground, the roof on which they had stood falling in before the last man had once again his feet on solid earth. Around the lumber yards on either side of Jones Falls, steam and smoke rose in such clouds that day was turned into night and firemen struggled along in practical darkness.

At length the united efforts of all the fire departments were beginning to tell and the final struggle for supremacy was short and decisive. A minor fire had been started by sparks in a woodyard across the falls, and for a moment it seemed as though past efforts were to be obliterated in this new development. But Baltimore and Chester faced it undismayed and human skill triumphed over its deadly enemy. From that time on it was a comparatively easy matter to confine the fire to the limits which it had already reached, and the last flames were extinguished towards the evening of that exhausting day.

New York long cherished a souvenir of the event in the shape of a stray dog which adopted engine "16" as its foster father and followed it faithfully through the streets all day. It accompanied the crew on their return and made itself perfectly at home in its new surroundings, responding to its name of "Baltimore" as though it had never known any other.

It is estimated that the temperature of the fire was rarely much in excess of 2,200 degrees Fahr., although in some spots it seems to have been approximately 2,800 degrees or more. According to various estimates the most intense heat in the fire-resistive buildings lasted from 30 to 60 minutes, varying with the amount of combustible contents, exposure and other features. Cast-iron radiators and typewriter frames were found in some places almost completely destroyed by oxidation, but had melted in a few cases only. Wired glass melted in a number of instances. In contra-distinction to ordinary fires in individual buildings, which usually spread vertically from floor to floor, this conflagration was essentially a horizontal fire as regards its attack and progress in each building. As a rule every story was ignited simultaneously through the exterior windows, and the fire swept across the building and out at the opposite side. Under such conditions the protection of floor openings will avail but little if the windows are unprotected.

Vaults made of brick walls built up from the ground, especially those having double walls with an air space between, made a remarkably good showing when provided with double iron doors, the outer ones being filled with about four inches of cement for insulation against heat. Vaults made of ordinary terra cotta tiles about five inches thick, and carried on the floors and structural frame, failed in a number of cases, owing to the fact that the tile was fragile and was cracked or broken by the heat. About 25 per cent. of the contents of the tile vaults was destroyed. Some of these tiled vaults also had double doors, each made of a single thickness of sheet steel with no insulation against heat. In a number of cases the inner door was left open, and the heat which radiated through the outer one destroyed the contents. Portable safes fared badly, approximately 65 per cent. of their contents having been destroyed. This was true of all makes of such safes, whether insulated with cushions of concrete or not.

It is a curious fact that the low bank buildings, on account apparently of their small height and in some cases sheltered position, usually escaped the maximum heat of the general conflagration and did not receive an extreme fire test. As a rule, they were partially wrecked by falling walls of higher buildings. A group of high office buildings of steel and terra cotta tile construction were typical of what may be expected from structures of this type, and it is interesting to note that the damage was generally greatest in the stories above the first. Notwithstanding the fact

BALTIMORE FIRE.   SKELETON STEEL CONSTRUCTION

that practically no water was used by the fire department in any of these buildings, the basements and, in some cases, the first stories were, to all intents and purposes, untouched, although the floors above were completely burnt out. Even the wooden nailing strips, which were embedded in cinder concrete below the top flooring, were entirely destroyed.

It was also specially noticeable that, although the conflagration attacked the fire-resistive buildings with great severity, the largest damage to the interiors was due to the fires in the buildings themselves. The damage was appreciably greatest where there had been a considerable amount of combustible material in storage. Even the severest injury to the exterior finish of the walls occurred over the windows on the leeward side, when the fire came from within.

Such was the great fire of Baltimore, the effects of which staggered the insurance companies of two continents and sent not a few into liquidation. But, as is often the case, in such events, it brought in its train fresh channels of thought anent fire control, while the energy and enterprise of its citizens has quickly obliterated all signs of the lamentable occurrence. Without going too deeply into problems which are dealt with in general elsewhere, there is one point that must make appeal to even the veriest tyro on fires and their fighting; namely, that Ovid when he penned the lines

> "Beginnings check;
> Too late is physic sought"

was giving the world in epitomized form the very key to the mastery of success against flames.

The writer must plead the indulgence of his readers if, in describing the great fire which destroyed the Equitable Building in New York, the narrative is related in the first person. Owing to the fact that he was so intimately associated with the events of that memorable occasion, to deal with it otherwise would be impossible, having due regard for the interests involved.

At 5:55 on the morning of January 9, 1912, the gong in my quarters struck 2-2-24, which indicated a second alarm from station 24, at the corner of Nassau and Pine Streets. Two minutes sufficed for me to cover the distance of about one and a half miles between my quarters and the scene of the outbreak, which proved to be in the Equitable Life Assurance Building. This was an oldish structure, eight stories in height and occupying the whole block, bounded on the north by Cedar Street, on the south by Pine Street, on the east by Nassau Street, and on the west by Broadway. The three first-mentioned thoroughfares were extremely narrow and contained buildings of considerable height, though some of them were of antique construction and doubtful fire resistance. On entering from Pine Street, I ascended the main stairway to the fourth floor, whence looking up I could see that a considerable area of the stories above was involved.

I immediately directed my first aide, Lieutenant Rankin, to send out a third alarm, and then proceeded to the fifth floor, where I met Acting Deputy-Chief Devanny, the officer in command previous to my arrival. Subordinate to him, and directing the companies, were Battalion Chiefs W. J. Walsh and George Kuss. One glance at the situation sufficed to impress me with the great battle ahead, and at once I ordered a fourth alarm with a special call for water tower No. 2. Water tower No. 1, which had responded on the first alarm, was already raised on the Pine Street side of the building. I returned to the street with a full grasp of the conditions to be met. A sixty-mile gale was blowing with the thermometer near zero. The direction of the wind was W. S. W., and I foresaw that it would drive the fire towards Nassau Street, where several old buildings, such as the Mutual Life and the Fourth National Bank, lay directly in its path. At this point Nassau Street is only 47 feet wide, and should the flames have swept the buildings to the east, under existing weather conditions, an uncontrollable conflagration would have resulted. To protect this point, therefore, was the first manœuvre and the reason for acquiring an additional water tower.

The second alarm assignment reported to me on my return to Pine Street and Broadway, and Acting Chief Kelly, of the 3rd Battalion, was immediately ordered to take command in Nassau Street. Engine companies were assigned to him and ordered to take their lines to the roof of the Fourth National Bank to drive the fire back when it broke through the eastern wall of the building, as was plainly evident would soon occur. Captain Henry, Supervising Engineer, was directed to meet water tower No. 2 on its arrival and have it placed in Nassau Street directly in line with the centre of the Equitable, connecting it with the high-pressure hydrants in Maiden Lane, and to order the high-pressure pumps started at a pressure of 200 pounds. This was done to reinforce the lines on the roof of the Fourth National Bank.

It may seem to the layman that the transmitting of the alarms, the assignment of companies and the hundreds of orders consequent thereon would take an appreciable length of time. Yet, from the moment the gong struck in my room until all arrangements had been perfected, exactly six minutes had elapsed. The actual plan of battle was evolved in less than 30 seconds after my arrival, and from that plan I never deviated.

Knowing the construction of the building, with its four entrances and corridors leading therefrom to a great central staircase, it seemed doubtful from the first whether the blaze could be conquered, but the motto of the department under my command has ever been, "Fight to a finish," and hence we endeavoured to outflank the fire by working from the staircase to windward, i. e., towards Broadway on the Cedar Street side of the building. Similar tactics were employed towards Nassau Street to confine the fire to the Pine Street side, between the streams directed by the twelve companies in the interior and the heavy volume of water from the lines placed on the upper floors of the buildings on the south of Pine Street. Such was the first line of attack, and a second line was at once provided by the companies in Nassau Street and the tower stationed there. It is my deliberate opinion that the interior dispositions of the forces at my disposal would certainly have been sufficient, and have succeeded in quelling the fire, while the regrettable loss of life which followed would have been avoided, had it not been for the criminal weakness of the iron columns supporting so heavy a roof as that which surmounted the Equitable Building.

The report of the New York Board of Underwriters on the subject is as follows

"The columns appear to have been very defective, due to the shifting of the core during casting, making one side of column very much thinner than the other. . . . Their condition indicates beyond much doubt that the initial collapse in each case was due to the failure of one or more cast iron columns."

Thoroughly mindful of this circumstance, I ordered every person but the firemen from the premises. At the moment there were hundreds of cleaners and other people within its walls, absolutely ignorant of any danger, as indeed to the ordinary observer there were no untoward signs, and only trained experts could detect the presence of peril. It is a matter of considerable difficulty to persuade persons, who fancy they have business, to leave their occupations and vacate their offices under such circumstances, and some time elapsed before the police reported to me that all but the firemen had been ejected. Alas, there were several who never obeyed the summons, as subsequent events were only too clearly to show. The fight now continued with increased persistence. I inspected Nassau and Cedar Streets, which, being to leeward, gave me some anxiety.

Returning to the Pine Street corner of Broadway, I watched for a few moments the battle which was being brilliantly fought. Never did men struggle harder or with greater intelligence; every order was promptly executed, but notwithstanding the stubborn attack from both within and without, I could see that the fire was slowly gaining. Until now my reports from inside had been favourable, but judging from external conditions I had grave doubts as to whether the officer in charge of those forces had correctly gauged the situation. It was this which determined me to make another inspection in order to satisfy myself. Accompanied by Lieutenant Rankin and Firemen Henry and Blessing, I proceeded to the 4th, 5th and 6th floors. Chief Walsh was in command on the 4th floor, and Chiefs Kuss and Devanny on those above. Chief Walsh was confident that he could drive the fire back and confine it to the Pine Street side. It must be emphasized that the conditions were good; very little smoke to weaken the men being observable, as it was driven eastward by the fierceness of the gale. Followed by my aides I returned to Pine Street, where I found that the granite trimmings on the dormer windows of the upper floors were beginning to fly. This told me at once of the intense heat which must be surrounding the unprotected iron columns of which mention has been made, and, in consequence, I ordered all companies to back down and out of the building.

A most critical stage of the fire had now been reached. I knew well that within a few minutes of the companies inside the building shutting off their streams the fire would gain complete mastery; hence the problem was to get the men out and into position with the second line of attack which had now become defense. It is an axiom of warfare that an advance is easier to conduct than a retreat. With such a furious and destructive enemy as fire, the task is even more hazardous. Thus it was obvious that the companies on the upper floors should go first, while their comrades on the lower floors, and in less exposed positions, held the flames in check and covered them with their streams. As soon as the latter were shut off the fire burst through with increased fury, but was met and checked by the lines in the surrounding buildings reinforced by the Nassau Street water tower.

As a further precaution the Pine Street tower was

moved to the corner of the former street, ready to en-
filade the fire and throw a complete water curtain across
Nassau Street in front of the Fourth National Bank,
should such a manœuvre become necessary. The time
was now 6.28 A.M., and I turned in a fifth alarm and
ordered an additional 25 pounds on the high-pressure
system. I had now 23 engine companies, 6 hook
and ladder companies, 2 water towers and a force of 275
officers and men. Lieutenant Rankin was despatched with
a second order that the men in the Equitable should back
down and get out with all speed, bringing their lines to the
Nassau and Cedar Street side, which was the quarter by far
the most dangerously exposed. All companies had now
reached the main staircase, except Engine Company 4 and
a few men from hook and ladder 1, who, under the direction of Battalion Chief Walsh, were fighting obstinately
and receding inch by inch. For the third time I sent an
order, adding that it was imperative that he and his men
should abandon their position, which had become untenable,
and leave their line.

Walsh received the message, but his sense of security, coupled with a desire to have one last bout with
his foe, caused the delay which brought him death. A
portion of the roof on the south side collapsed, forcing out
part of the wall of the inner court and burying the steps
down which the last men were hurrying. Before this catastrophe had occurred the companies who had responded on
the fifth alarm had been assigned positions in the buildings
on the east and north, where they connected to the stand-
pipes and threw powerful streams into the upper floors of
the burning structure. Their efforts were successful, and
at this point the flames were held in check.

Captain Farley, of Hook and Ladder 8, now
reported to me that he and his men had removed
Captain Bass and some members of Engine Company 4 and Hook and Ladder 1 from the collapsed
part of the building. I then ordered a roll-call and
discovered that Battalion Chief Walsh was missing. A

EQUITABLE FIRE, CORNER OF BROADWAY AND CEDAR STREET.

search party was instituted to rescue him, but failed in the attempt. I learned that as Chief Walsh was about to descend the stairs to the third floor the unmistakable rumble of falling walls warned him of his danger. Had he remained where he was he would have been unscathed, but he sprang over the rail and dashed towards a door on the left leading to Nassau Street. Could he have reached it he would have cheated fate, as did two of his comrades; but he was buried in the wreckage within two feet of safety. Never have I known a man more enthusiastically devoted to his calling; it was the breath of his existence. Brave as a lion, and loving a fight for its own sake, he constantly studied to increase his technical skill. It was my knowledge of the man, of his bulldog grit and determination to conquer or die, that caused me to be so insistent in my commands to him to leave his post of danger. His heroic spirit was shown in his last action. Aware of the peril, he called to Captain Bass, of Engine Company 4, "Go at once. Save yourself and your men," and he remained to add one more name to the roll of those who have died nobly in harness.

It was now that the full force of the millions of gallons of water began to tell. The water tower in Nassau Street was sending forth a heavy stream through the two-inch mast nozzle at a pressure of 120 pounds to the square inch, supplied by the high-pressure main in Maiden Lane. This was directed against the flames roaring through the Lawyers' Club on the fourth floor, while from the roof of the bank across Nassau Street, Acting Battalion Chief Kelly was performing admirable service. On the south, Battalion Chief Rush had availed himself of the stand-pipes in the buildings and was using our steamers in conjunction with the house pumps, thus being able to obtain a considerable pressure. The same plan was carried into effect in Cedar Street, and every exposed point was covered on all sides. I was congratulating myself that we were masters of the situation when the fury of the gale increased. Gusts of wind, attaining a velocity of seventy miles an hour, swept across the open space formed by Trinity graveyard at the southwest side of Broadway, and the mercury fell steadily and remorselessly. So intense became the cold that dripping walls turned to ice and the streets were frozen lakes, while enormous volumes of water were turned to spray by the wind a few feet from the nozzle. Men were repeatedly thrown down in their efforts to cross the path of this hurricane, and I myself was taken from my feet not once but twenty times, and dashed against the wall of the building where I stood. The Equitable now resembled a volcano in eruption. Great masses of granite from its walls were being tossed high in the air like thistledown and exploding a hundred feet above our heads from the intense heat, their fragments falling in meteoric showers about us. A great section of the outer wall burst near the corner of Pine Street and Broadway, and a piece of stone weighing several tons fell near Mr. Robert Mainzer, with whom I had been speaking, missing him by only a few inches. I then closed that side of Pine Street, even forbidding firemen to pass along it. The intense cold seemed to give the flames a peculiar glow, while the high wind spread them fanwise, flickering and beckoning over the ice-bound streets. There comes a time in a fire of this description, which marks the beginning of the end. If outside exposures are properly protected there can be no possibility of any increase in the conflagration, and it will be confined to the smallest possible space. Then one of two things will occur; either the contents of the building will burn out, leaving no food for the flames; or it will fall. Should this latter contingency seem imminent, men must be kept at a safe distance from the walls, and judgment must be used to determine what is the limit of danger. Sometimes a wall will fall outwards at full length as though on hinges, covering the width of a street; then again it will collapse, break in the middle and fall in a heap like a house of cards. Needless to say, the first of these two conditions is the most dangerous in all respects, and must be guarded against at any hazard. In the event of a simple collapse the fire has then passed the crisis, and as soon as this occurs men can immediately be advanced to close range without special danger. The roof and floors of the Equitable Building were heavy, and the intensity of the heat was so great that I feared it would expand and force the outer walls. Under such weather conditions as existed, and in the narrow streets, this would have been a serious matter, and every nerve was strained to its utmost to drive the fire back and to hold it in the centre of the building. This attempt was crowned with success, due not only to the powerful apparatus at my disposal, but to the intelligent and, in many cases, brilliant operations of both officers and men.

A fire chief can never tell what may happen from one minute to the next, and fires bring many surprises in their train which call for quick action of mind and body on the part of the officer in command. This day was to prove no exception to the rule. Just as I felt that the fight was won, and was expecting an inward collapse of the floors on the Broadway side of the building, word was brought that three men were on the roof overlooking that street and calling piteously for help. After all my efforts to clear the building it seemed impossible that anyone could have remained within its precincts. And yet these poor cleaners and porters had defied a command and had pitted their judgment against scientific knowledge, with the result that they had been driven to the roof, where we could see them standing. To reach them on that spot over 100 feet from the ground, when the possibility of a collapse had become imminent, was a task to test the nerves of the strongest and the bravest of men. That an attempt at rescue was fraught with great danger to all concerned I had not the slightest doubt, but it is the duty of men on such occasions to brave death and even to defy it. All chances were against them. Momentarily I was expecting an avalanche of bricks, stone and burning embers; the fierce gale swept strong men from their feet, and the spray from the nozzles froze on their faces until they could scarcely see. In spite of these conditions the men responded to my call without hesitation. A hook and ladder truck was swung in on the northwest corner of Cedar Street, and in less than one minute the extension ladder had been raised. As I stood at its foot I did not have a chance to ask for volunteers or to order any men to this terrible duty. On the instant Lieutenant Rankin, Firemen Molloy and Blessing sprang on the rungs, taking scaling ladders with them. In the meantime I could see that it would be a most difficult undertaking to scale the Equitable Building on account of the projecting cornice, and I therefore ordered Acting Chief Kelly and the officers and men of Hook and Ladder 1 to proceed to the ninth floor of the building on the north side of Cedar Street with the gun, roof rope and life line. If the line could be shot true against such a gale it might serve two purposes, for it would be ready for use by the men ascending the ladder, or, if this attempt failed, the captives could make the line fast to some projection and slide down to possible safety. The shot was aimed and the line fell true. We could see the men in the act of hauling it across the space when the expected happened. The great collapse came. With a cry of agony and despair the unfortunates sprang out into the air, and as they plunged downwards there came with them the roof and upper floors. From my position I at first thought the bodies were those of the brave fellows who had so nobly gone to the rescue, and though they struck the street a few feet from where I stood, and though fire, smoke and débris were on all sides, for an instant I felt indifferent to my own fate. Then I realized that other lives and vast treasures were at stake, and that at this moment my life was of value to the city.

Turning around, I walked slowly to the centre of Broadway, and from this point I could see that the men who had ascended the ladder were alive. Blessing was on the ladder, Rankin had one foot on the ledge, and Molloy was standing on the highest ledge of the broken and badly bulging wall. Their efforts had been in vain, but heroism could have been put to no greater test. My relief was great when I saw them descend unhurt. And now horror succeeded horror with incredible rapidity.

EQUITABLE BUILDING, CORNER OF BROADWAY, WHERE THREE MEN JUMPED TO DEATH.

Scarcely had the unfortunate creatures who had jumped from the roof been .removed from the street, when Fire Commissioner Johnson told me that he had been informed that there were men · imprisoned in the .vaults on the Broadway side. The windows of these vaults were protected by bars of iron two inches in thickness, and were inset at such close intervals that no human body could possibly pass between them. I soon found that this information was correct, for there, caught in a fiery prison, were three men; two living, pinned down by broken floor joists, and one dead, killed by a falling beam. With a raging fire behind them, a raging fire over them, heavy iron bars in front, and broken and tottering walls on every side, their predicament was a terrible one.

"Save them, save them," was the cry from men who stood at a distance. But this seemed to be impossible. I directed two companies with sledges and other heavy tools to try and wrench the bars. In addition, though scarcely to be mentioned in comparison with these precious lives, there was a billion dollars' worth of security in the vaults, and the fire threatened both with speedy destruction. Fully realizing the gravity of the conditions, and wishing to obtain a better view of the situation, I took the elevator to the eighteenth floor of the Trinity Building directly across Broadway. When I reached the front window overlooking the Equitable an awe-inspiring scene met my gaze. Beneath me lay a seething, boiling cauldron. The very earth seemed to vomit forth flames and send up from its depths mammoth tongues of fire. Parts of chairs, desks and boards were being hurled like pebbles five hundred feet into the air. Only the pen of Dante or the brush of Verestchagin could do it justice. But the question for me to decide was whether the Broadway front would hold, or whether it would collapse burying the entombed men and the companies trying to effect their rescue. After careful survey I determined that the walls would stand, but to ensure this I ordered that a strong stream of water from the Trinity

Building be employed to reduce the expansion by forcing the fire back at this point.

About this time I resolved to transmit the borough call, feeling that additional aid was necessary properly to protect the vaults and the men imprisoned therein. It was also advisable to have a greater number of powerful streams on the leeward side of the fire, although up to this time I had been able to hold it in check. Now the time seemed to have arrived for an advance, and this my lines were unable to accomplish. All these conditions made the borough call a necessity, and the alarm 7-7-24-3-3-39 was transmitted. Translated into plain English this meant that the companies assigned to respond on the third alarm to Box 39, Borough of Brooklyn, would proceed to Box 24, Borough of Manhattan.

The Brooklyn companies arrived promptly in charge of Deputy Chief Lally, and were assigned to positions with the exception of the water tower, which was not needed and was sent back to quarters. Two engines were connected to the siamese inlet on the front of the Trinity Building and 2½-inch lines of hose attached to the stand-pipe outlets on the 7th, 8th, 9th and 10th floors. These were all stretched to the 8th floor and connected in pairs by means of 2½- to 3-inch siamese, then a length of 3-inch hose was connected to each of these, and in turn to a 3-inch siamese. Leading from this was a length of 3-inch hose having a 1⅜-inch nozzle. This provided a pressure of 130 pounds at the nozzle, with 260 pounds on each engine, and had less friction than if any other method had been employed.

Now began the battle for life and treasure. Hack-saws were procured, and for almost an hour Engineer Larke assisted by Rankin, Henry and others, sawed at the bars, while great masses of stone fell from the upper stories around the workers. One great fragment rebounded and struck Larke in the back, almost paralyzing him. Rankin now took the hack-saw and cut through the remaining bars, so that ropes could be attached and the opening sufficiently enlarged to admit of the passage of a body.

One of the men was taken out suffering from smoke, exposure and shock. But the other cried, "For God's sake don't leave me, my arm is fast!" Upon examination, it was found that his arm was pinned across the back of the dead man by two iron beams, and for 15 minutes Henry and his comrades, using crowbars, pried and pulled, assisted by the man himself, before his release was effected. When free he collapsed, and was taken across Broadway, where he joined his companion under the care of Dr. H. M. Archer, who gave them every attention that humanity and science could suggest.

By this time the fire was well under control on the north and east, and all danger of its crossing Cedar and Nassau Streets had passed. I now called a boat tender and stretched 3½-inch hose from the high-pressure hydrants to the Broadway front of the building. Three-inch lines were also taken from the water tower into the Cedar Street buildings opposite the vaults and company lines were siamesed in order that heavier streams could be forced against the gale which still increased. This method was in operation on the roof of the Clearing House, where it was most effective.

A peculiar phenomenon of this fire was that it worked steadily to windward against the furious gale, and it seemed as though determined to destroy the enormous wealth contained in the vaults. All our forces were now concentrated to prevent such a catastrophe and also to prevent the cremation of the lifeless companion of the two men we had rescued. Owing to the magnitude of our attack the securities were untouched and unharmed, the walls, which were badly cracked and out of plumb, remained standing, and the corpse was not incinerated. Cautiously we now closed in and the fight was over. There is one incident which I must mention, as it serves to show the hold sport maintains upon its votaries even in moments of the greatest strain. Mr. August Belmont came to me and asked permission to go through his offices, which, facing the east had to a large extent escaped the great damage experienced elsewhere.

I personally went with him through the ruins of his once beautiful suite of business premises, now sadly spoiled by water and fire. He then explained to me that his chief fear was lest harm should have come to the records and pedigrees of his horses, which of course are famous not only in America but wherever racing is popular. I am happy to say that I found them intact, and with a smile he tucked them under his arm and bade me a cheery good-day.

Another fact which I take pride in recalling was that of the gallantry shown by Father Joseph P. Dineen, who, at the risk of his life, conveyed the Sacrament to one of the men afterwards rescued from the vault, this at a moment when all onlookers feared the worst, and no man would have been considered a coward for hesitating. Some idea of the magnitude of the operations can be gleaned from the following statistics. Eighty-five officers and about five hundred men operated thirty-one steam engines, ten hook and ladder trucks, two water towers, and superintended the high pressure service, while the water used in the attack amounted approximately to twelve million gallons. During the progress of the fire, all business in Wall street was suspended and anxiety reigned in two continents as to the fate of the billion dollars' worth of securities in the strong rooms. It speaks volumes for the skill of the fire-fighters that not one dollar's worth of damage was done in that direction and that when recovered the papers were not even discoloured. The outstanding features of this remarkable fire were the tremendous value of the property at stake, the extraordinary climatic conditions, and the possibility of the spread of the flames, which would have caused a disaster unparalleled in the annals of history, so stupendous would have been the financial loss.

In addition to this, the construction of the building concerned was something of a revelation to all thinking persons; for the weakness of the columns supporting the roof was so glaringly apparent even to the lay mind that those

ENGINE AT EQUITABLE FIRE.

responsible for its erection must have been either hopelessly incompetent or criminally careless. Further, owing to the age of the structure and the idiosyncrasies of some of its tenants, it was a literal rabbit warren of private staircases ending in *cul-de-sacs* and narrow passages leading nowhere in particular. The only marvel, in fact, appears to have been that the loss of life was not greater, for it was only too easy for the firemen operating to lose their way in the intricacy of its mazes.

I must not fail to compliment the Police Department on its excellent work in the keeping of the "Fire Lines," this work being exceedingly difficult, owing to the extreme exposure and extraordinary conditions prevailing.

It is pleasant to record that financiers and others of wealth and prominence with offices adjacent sufficiently recognized the self-sacrificing devotion of the Department in subscribing the sum of $185,000, the interest of which was to be used in perpetuity for the benefit of widows and orphans of firemen and policemen killed in the discharge of their duty.

CHAPTER XI

PART II

THERE occur at intervals in the history of the world, ca lamities occasioned partially by fire of which it is almost impossible to give a concise narrative, or upon which either to pass criticism or apportion blame. In other words, when fate or destiny, or call it what you will, takes a hand in the game, human ingenuity, science, and forethought can only play subsidiary rôles in dealing the cards. The Baltimore fire was destructive of property, the Equitable teemed with terrible possibilities and gave scope for the most modern fire strategy that probably the world has yet seen, but the conflagration in San Francisco formed an upheaval of primal elements which, in their magnitude, stand alone in history, and yet show that dogged perseverance inborn in the fire-fighter, which sooner or later surmounts the greatest obstacles.

On April 17, 1906, San Francisco was one of the happiest, grandest, most popular of cities in the United States. Within twelve hours a large portion was in ruins, within twenty-four it was a mass of belching flames, and within thirty-six the lamentations of its inhabitants had penetrated to the most remote quarters of the globe. To epitomize this ghastly débâcle. On Wednesday, April 18th, an earthquake shock occurred, doing considerable damage, so badly crippling the water mains that, though their supply was rated at 36,000,000 gallons a day, not only was the fire department unable to obtain the wherewithal with which to attack the ensuing fires, but so scarce became this necessity of human life that it is credibly reported that at one period it was being retailed to thirsty thousands at fabulous sums per cup. This conflagration destroyed 2,831 acres of business and dwelling houses, and caused losses to the insurance companies concerned of approximately 300,-000,000 dollars. Needless to say, it is impossible to determine the number, location or causes of the original outbreaks. All that can be definitely stated is that the fire alarms at headquarters were completely dislocated by the earthquake shocks, that the building in question was subsequently burned, that the telephone service became completely disorganized and that doubtless many unsuccessful attempts were made to apprise the fire department of the need of its services. All that can be hazarded is that, within half an hour of the commencements of the outbreaks, there must have been twelve distinct and separate fires needing attention. Roughly, in order to give some idea of the operations involved, it may be stated that the centre of the fire zone was an eminence known as Nob Hill. Thence one portion of the city was involved eastwards to the waterfront, taking in Chinatown and the Latin Quarter en route; a second spread in a southwesterly direction through the business section and menaced the wharves and ferries; while the third, originating in the Mission district to the west of Nob Hill, burnt its way steadily towards the Union Iron Works, where at that time were building two battleships for the United States Navy. Before dealing in such detail as is possible with the incidental operations of the fire department, it may be said that the fire force, including reserves, consisted of some 600 men, 53 engines, 15 ladder trucks, 9 chemical engines and 2 fire-boats maintained by the Harbour Commissioners. One of the fire-boats had a capacity of 1,400 gallons per minute and the other 930, both with a water pressure of 150 pounds. Of the 77,000 feet of leading hose, nearly 38,000 feet were lost, or over one-half; while 3 engines and a ladder were disabled beyond repair. Fire Chief Sullivan was unfortunately injured at the outset and died before he had formulated a plan of attack. This comprehends the total casualities to men and material in the department during the entire conflagration, a remarkably small percentage of the whole, and it is a fair supposition that had the means of regular communication been open and had water been obtainable during the early hours of the disaster, and having also due regard for the lightness of the wind and its direction, the fire department would have obtained control by noon of the first day.

During the first period, that is to say until Wednesday night, the fire appears to have been spasmodic and did not possess the nature of a fierce, sweeping blast. The ordinary rules of exposure seemed to have prevailed, and a leading part was played by familiar factors, such as individual combustibility, adjacency, opposing openings, short distances and excess height. Some notable cases of defense are worthy of comment, such as that of the U. S. Mint, an old building far behind modern standards of fire resistance. Superintendent Leach, of the fire department, rallied his men and, assisted by some regular soldiers, beat the fire off in a manner worthy of the highest commendation. Another remarkable effort was that made by the employees of the Post Office to save that structure. As the flames attacked through windows broken by the heat, everything igniting was extinguished in detail. The officials fought most gallantly, and three days later, when it was possible once again to obtain access to the building, eleven postal clerks, who had been seventy-two hours without food or water, were rescued, together with the whole of the mail of which they had been in charge.

Late in the afternoon, the great twenty-one story Spreckels Building ignited, through broken windows on the fourth floor, from fires started in two small frame buildings adjacent to it. This provided one of the most spectacular scenes of the whole outbreak. Enormous crowds watched the dull red glow mount floor by floor, till it reached the ornate three-tiered dome surmounting this edifice. The circular windows therein seemed to shine like moons for some moments, then followed a thousand spurts of flame as the floors collapsed, and, as darkness closed around, men and women wailed hysterically thus to see the pride of their city so remorselessly destroyed. As for the Palace Hotel, its doom was sealed early in the afternoon. A fine attempt was made by its staff, assisted by some of the guests to resist the enemy, but the protection of a hundred odd closely attacked and wooden-framed windows and a vulnerable roof swamped them, and the hotel was abandoned.

Shortly after this commenced the extensive use of explosives which figured so prominently in this conflagration. It is not surprising that men reduced to helplessness and desperation by lack of water should have resorted to what has been proved in all modern fires to be useless, and, in the opinion of the writer, even harmful. As is usually the case, the explosions made no effective gaps and rather served to increase the quantity of combustible material. On the other hand, windows throughout the neighbourhood were shattered, the proximity of exploding buildings made it dangerous for owners to prosecute individual efforts towards the protection of their own property, and it would appear that the choice of location for this desperate expedient was both haphazard and unintelligent. The situation when Wednesday night arrived is important to realize. Until now the rich business district, north of Market Street, and the high-class residential area were untouched. It was still possible to maintain communication and to conduct organized opposition, since the centre of the city was yet habitable. But human nature had become exhausted; questions of life became paramount to those of property, so that upon the direction of the wind depended the future. Alas, during the evening the breeze, for it was little more, veered southward and increased just sufficiently to level the sweep of flame and render leeward positions untenable. The huge frame of the Mechanics Pavilion was transformed into a roaring pyre and the upslope towards Russian Hill perceptibly in-

creased the vulnerability of the district. From now onwards the spread of the flames was more rapid, and they
greedily ate their way along O'Farrell Street devouring in
turn theatres, hotels, clubs, stores and apartment houses.
Higher buildings, like the Crocker, felt the blast of the intense heat in their upper stories and caught fire ahead of
their time. Fireproof buildings, like the Mills and the Merchants' Exchange, which during the day had proved bulwarks of safety, became involved and towards midnight
were burning like beacon flares.

A most desperate stand was now made around the Fair-
mount Hotel. Sailors from a revenue cutter, assisted by
firemen, ran a three-quarter mile length of hose from their
ship to the building, their officers with drawn revolvers im-
pressing civilian bystanders to act as property savers. But
all to no purpose, and as the dawn of the second day coloured the eastern horizon it was realized that not only the
hotel but all the surrounding wealthy residences were
doomed. During that Thursday morning the wind light-
ened, and now blew from the east and served to check the
advance of the flames which threatened the Ferry Building.
It confronted, however, the defenders with a fresh and even
more alarming development—that of losing the only closely
inhabited part of the city remaining—the section west of
Van Ness Avenue. In this 125-foot street, the most extraordinary efforts had been resorted to, in a vain attempt
to stop the ever spreading fires. Beautiful houses were
blown to atoms by dynamite, while the artillery, belonging
to the military garrison, had carried on a steady and re-
morseless bombardment with high explosive shells. The
neighbourhood was an inferno; above the crackling of the
flames resounded the dull boom of bursting shrapnel and
the cries of terror-stricken men and women, while a canopy
of green-gray smoke slowly spread upwards marking the
positions of the targets. Yet all this only served to provide
fresh fuel for the oncoming conflagration. Some check was
doubtless afforded by these drastic measures, but the invader still advanced westward. On the Friday morning,

PROGRESS OF THE SAN FRANCISCO FIRE: FIRST PERIOD. FROM BROADWAY AND TAYLOR STREET LOOKING SOUTHEAST. HALL OF JUSTICE BETWEEN SHOCK AND FIRE.

the third day of the fire, the east wind happily dropped and was succeeded by a strong westerly breeze which, within the course of a few hours, shifted between northwest and southwest, the former driving the flames into the Latin Quarter and destroying the frame houses comprising it like so many dry leaves, and the remarkable sight was witnessed of thousands of barrels of wine being stove in with the vain hope that the liquor might be used to stay the approaching cataclysm. Forces concentrated near the Merchants' Ice and Cold Storage Co., with the assistance of a city engine, and using the company's own water supply, at this point won a victory over the flames. Individual work also saved an isolated and somewhat scattered group of high class dwellings on the precipitous summit of Russian Hill. The conflagration had thus lasted three days, and on the Saturday morning a heavy rain did much to bring the situation under control. A few smouldering blazes along the east water front occasionally flared up, endangering unburnt structures, but were, however, promptly suppressed. Vigorous and effective measures were now taken to prevent new outbreaks in the uninjured districts where, owing to the earthquake, chimneys, gaspipes and electric wiring were generally in an unsafe condition, and where the scant water supply rendered the situation most precarious. No time was lost in destroying dangerous walls, and it is worthy of comment that explosives were again used to an exceptional degree in this work, causing unnecessary additional damage in some places and, unfortunately, quickly terminating many opportunities for distinguishing the true effects of the fire. Thus, within the burnt area of 2,831 acres there survived in a partially habitable condition: firstly, three groups of buildings, i. e., the detached dwellings on Russian Hill, some warehouses at the foot of Telegraph Hill and a mercantile group near the Custom House. Secondly, one factory plant—the Western Electric Company. Thirdly, three government buildings—the Mint, the Post Office and the Appraisers Building. Fourthly, two fire-resisting office buildings—the Hayward, with a three-story building adjoining, and the Atlas Building, with a two-story structure adjacent to it.

Such is a brief description of the conflagration which devastated San Francisco and necessitated without exaggeration the foundation of a new city. The narrative has been shorn of anything that might detract from a realization of the factors which governed the actual situation, though naturally it goes without saying that incidents of interest, humorous, pathetic and tragic abound.

As in all great crises, the behaviour of those concerned varied according to temperament and circumstance, but, generally speaking, there was little real panic, and on all sides was observable a tendency to make the best of things and incidentally to help others to do likewise. At first people were so stunned that they scarcely realized what was passing, as was evidenced by one stranger to the town, who, making his way to safety, was accosted by a rough who demanded his purse. He surrendered it without demur, but the hold-up had been observed by an officer in command of some soldiers. Martial law having been declared, the thief was shot dead on sight. Afterwards, being asked to give evidence regarding the shooting, the victim of the assault was found to have forgotten everything about it and remarked that he was so bewildered that anything seemed quite natural. This curious mental effect was by no means uncommon, and no doubt indirectly exerted an influence against any access of unreasoning and overwhelming terror which would have rendered the exertions of the authorities practically abortive. A story, dramatic in its sheer horror, was related by a doctor who reported that he had found a man pinned under débris and suffering the most horrible torture, the while calling loudly for some one to put him out of his misery. After consultation, a police officer drew his revolver and fired at the sufferer, but being presumably unnerved, the shot went wide of its mark. The doctor was then authorized to act, and he accordingly opened the arteries in the man's arm, thus assuring him a speedy release from his agony. Thieves there were, too, in plenty, though short shrift was their lot when caught. Firing squads patrolled the streets, and these ghouls paid the price of their hideous crimes, the hacking of beringed fingers from lifeless hands and the like, with their own worthless bodies. On the other hand, simple heroism could be depicted in no nobler form than the spiritual comfort extended to the dying by the ministers of all denominations, who worked like slaves at great risk to themselves.

A word of praise must be written anent the pluck and never flagging determination shown by all ranks of the fire department under the command of Chief Shaughnessey, who succeeded Chief Sullivan after the death of the latter. The firemen worked for three whole days with such apparatus as was at hand, and only ceased when compelled so to do from physical exhaustion. And, withal, humour was not lacking. It so happened that the Metropolitan Opera Company, of New York, was fulfilling an engagement in the city at the time, and the experiences of its individual members would fill a volume. Their worldwide fame, of course, aroused the greatest interest in their fates, and it was only after some days that public anxiety was allayed and it was learnt that no one of their number was the worse for the experience. Caruso was a guest at the Palace Hotel and only escaped with difficulty. But he accepted the unexpected with a philosophy not usually associated with his countrymen and, as he sat in the middle of the street upon his valise wondering what was coming next, he nonchalantly rolled a cigarette and professed himself as not unduly disturbed. Later, in common with everyone else, he was compelled to shift for himself, and owed his cordial reception by a band of soldiers, who gave him food and lodging, to the fact that he was carrying with him a photo of ex-President Roosevelt inscribed with the words, "With kindest regards." This served as a passport, one of the men remarking, "If you're Teddy's friend come right in and be comfortable." Caruso afterwards summed up his impressions in the sentence, "It instantly recalled the horrors of my native Naples, of which I've been reading. Vesuvius in eruption could not have been as horrible." Campanari, the great baritone, contented himself by opining that it made a change in the monotony of touring, and that he found Caruso's pajamas, in which incidentally he had escaped, a bad fit. Rossi, the bass, passed the time by trying his voice; while Nahan Franko, one of the conductors, risked his life by returning to his hotel in order to save a violin he much prized. Madame Sembrich succeeded in saving her pearls, reputed to be some of the finest extant, but assessed the loss of her wardrobe at $25,000. Finally, Alfred Hertz, the musical director, who also helped himself to Caruso's garments in the moment of emergency, found safety near the zoölogical gardens, which, owing to the roars of the frightened beasts, he declared to be a more horrible place than any in the city.

A fact of more than passing interest, which must strike all observers, is the similarity of the results recorded in this conflagration to those in the Baltimore outbreak. The latter was the first in which modern methods of fire resistance received a severe test. There, the water supply was adequate and the Fire Department well up to the average and manipulated with considerable intelligence. There were fireproof buildings, most of them of modern construction, and so situated as to reinforce each other and act, so to speak, as fire-breaks. Yet, the result shewed that in the direct sweep of the fire as determined by the direction of the wind, nothing survived except the following: Firstly, an occasional one or two-story building favourably located as to shelter or wind currents. Secondly, an occasional grade floor in a fire-resistive building and, thirdly, the empty shells of the fireproof buildings themselves, none of which possessed front window protection. Finally, structures on the side borders of the wind sweep, where the exposure was confined to ignition from brands, and where men and apparatus could maintain a working basis and keep open their communications. There was also something in the nature of a successful check at Jones Falls, a stream of water of but moderate width, by which engines belonging to the New York Fire Department made a determined stand. Thus, from past experiences, there was no reasonable expectation in San Francisco of the survival of any building after the Fire Department was in retreat except in cases analogous to those just mentioned. In the main this proved correct, with some few exceptions. Within the burned section not only did all frame buildings succumb but also all brick structures having wooden floor beams, whether of good, bad or indifferent construction, and with more or less complete ruin in nearly every case with the one exception of the Palace Hotel.

Prominent amongst conclusions which may be formed from this disaster, in the opinion of the writer, are the uselessness of explosives as a deterrent measure to the spread of flames and the danger to tall buildings from the heat engendered by burning structures of a lesser height. The former accentuates confusion, causes panic, fosters misunderstandings between municipal and federal authorities, destroys property which otherwise might conceivably be saved, provides fresh fuel for the flames and hence is practically worthless as a serious feature in fire-fighting. An exception, which may occur, only goes to emphasize the point. As regards the latter, this danger was plainly exemplified in the occurrences in San Francisco and serves to illustrate the care which must be taken in considering the fire-resisting methods which must receive attention in the modern skyscraper, and which are dealt with at length in another chapter.

Suffice it to say that the heat wave generated during the climax of the conflagration rose to a height of about 300 feet above the street level and was directly responsible for the ignition of church steeples, skyscrapers and all structures of a similar character. Otherwise many old data received confirmation, which have been listed as follows in the Underwriter's report upon the conflagration.

A. The dangerous effect of a number of simultaneous fires.

B.   The weakening of a fire-fighting force if compelled to thin out over a wide front.

C.   The improbability, with existing methods, of frontal resistance to a fire sweep, when the wind velocity exceeds a certain critical figure.

D.   The special vulnerability of leeward upslopes.

E.   The structural ruin in conflagrations of all wooden joist brick buildings where the stability of the walls in any way depends upon the bracing of the beams.

F.   The limited utility in a conflagration of rear and side shuttering, where front windows remain unprotected.

G.   The likelihood of ignition of ordinary roofs, consisting as they do of wooden boards with a thin veneer of tin or other roofing material.

H.   The slight value as conflagration breaks of fireproof buildings, when abandoned.

I.   The possibility of holding buildings even with unprotected openings, provided there are some men, even only a little water and the openings are few.

J.   The structural survival, even without window protection, and when abandoned, of steel frame buildings with fireproof floor arches, provided the steel frame is properly encased with fireproof material, the structural damage being in close proportion to the quality of the frame protection.

K.   The greater or lesser destruction in such buildings of all non-structural interior; heavy spalling of all kinds of facing stone, the injury to ornamental mouldings and copings, extensive damage to hollow tile in floor arches and partitions as usually constructed, a marked increase of injury where wood finished floors are used over the floor arches, the danger from falling safes where there is loose back filling, the failure of unprotected cast-iron mullions and spandrels in courts and the weakness of roofs carried on unprotected steel rafters with suspended ceilings.

Amongst other important lessons derived from this conflagration in the matter of fire-fighting may appropriately be noticed the following:

A. The importance of front as well as rear and side window protection, fire-resistant if possible, but at any rate, fire-retardant, i. e., wireglass.

B. The necessity of encouraging individual protection by occupants of buildings.

C. The importance of ample water supply and good pressure.

D. The necessity for all fire departments to have a large reserve of apparatus and hose.

E. The importance to fire departments of powerful apparatus with long range.

F. The importance of fire-resisting roofs, roof structures and of well protected skylights.

G. The necessity of the adoption of rigid standards for column protection.

H. The importance of good bricklaying and mortar, with cement in place of lime.

J. The importance of efficient protection to the steel frames in roof attics.

K. The importance in partitions of a better bracing of tile and the need of fire-retardant transoms as well as doors.

In conclusion, perhaps the writer may be pardoned for hazarding the belief that in case of a great conflagration, where the military authorities are invited to assist in the maintenance of order, every effort should be made to assist the Fire Department, and the loss of individual property should be subordinated to the public weal, in accordance with the expressed opinion of the Fire Chief. Thus the policy at San Francisco, by which looting was prevented on any large scale by the indiscriminate employment of the military who were also responsible for the use of explosives, may have saved some thousands of dollars, but this very policy was probably accountable for the loss of millions, by the way in which the skilled fire-fighters were hampered in their movements through official interference, by the unnecessary blocking of important thoroughfares and by the fears of bodily harm consequent upon unexpected explosions.

It would appear as though the American continent possessed a monopoly of great conflagrations and, in all truth, this is in a measure correct, owing to peculiarities of construction. But Canada supplies an instance of what may happen when the fire department is not equal to the needs of the situation, which must sometimes occur when the building material is chiefly wood. The town of Hull, which is situated on the north bank of the Ottawa River, directly opposite the capital of the Dominion, was, until April 26, 1900, a thriving and prosperous municipality. On that spring morning a fire broke out a quarter of a mile from the main street of the little city and, fanned by a fierce gale from the northwest, rapidly advanced in the direction of the countless lumber mills and other factories from which Hull obtained its prosperity. The population was chiefly composed of persons employed in these industries and of the heads of the mills in the district, whose houses, although many of them large, were built of wood. By 11.30 the flames had swept across Main Street, and its dozens of cross thoroughfares were rendered impassable. The Court House, the Post Office and many churches were destroyed, and by midnight the interprovincial bridge, connecting Hull with Ottawa, was a mass of flame. In the ruins of Hull there remained only the Catholic Cathedral with a few houses clustered about it, and two factories to mark the existence of what had once been a flourishing industrial centre.

But the flames were unsatisfied. Aided by the wind, great masses of burning embers ignited the power houses, street electric and incandescent electric companies' buildings on Victoria Island, from whence the wharves on Chaudiere Flats, part of Ottawa itself, were within easy distance. Here were situated a great number of lumber mills, and the piles of dried timber were the most enticing food for the roaring conflagration that could have been found. Here also, was located the Canadian Pacific Railway station, which, being of wood like the other structures, offered no resistance to the attack. In fact, so rapid was the onrush of the enemy that many fine houses were consumed in the twinkling of an eye, and before their owners were able to save even the smallest proportion of their possessions. Montreal and smaller towns in the vicinity of the threatened city nobly responded with men and apparatus on an appeal for aid, since the outbreak had assumed proportions far beyond the control of a comparatively small local fire department. But even this assistance, combined with the efforts of the militia, proved of no avail in the face of the tornado of flame, which tore like a whirlwind past every obstruction and threatened to transform the Capital of Canada into a heap of ashes like its suburb of Hull. Rochesterville, a small township, which had been included some time previously within the city limits, was rendered a desolate waste, and had it not been that the direction of the wind mercifully changed to the east, and had it not been for the high cliffs which formed an insurmountable barrier to the onset, not all the fire departments in Canada could have saved the city. Owing to the destruction of the electric light supply, the House of Commons, which was then sitting, was obliged to adjourn.

Everything possible was done to provide shelter and subsistence for the seven thousand homeless people, whose condition was piteous in the extreme. Most of them were labourers from the mills, and lumber yards who had seen their homes wiped out and their occupations taken from them at practically one and the same moment. The military drill hall and the exhibition buildings were devoted to this charitable purpose, and many philanthropists proved themselves worthy of the demands made upon them. A curious feature of this disaster was the fact that, after the fire had burnt itself out, there remained no smouldering embers and smoking ruins, but all was literally in ashes, so thoroughly had the flames done their work. It is also worthy of note that only seven per-

sons met their death, and that no fireman was injured with
the exception of the Chief of the Hull brigade. The property loss was assessed at $17,000,000 (£3,400,000),
and some idea of the extent of damage in the
lumber yards alone can be gained from the bare statement
that two hundred million feet of timber was destroyed.
Needless to say, the price of this commodity was materially
increased and the trade suffered severely. This conflagration, it will be observed, was of the same sweeping character
as that of Baltimore, though fought under totally different
circumstances.

For sheer horror, the disaster at the bazaar in the Rue
Jean Goujon, Paris, on May 4, 1897, surpasses the wildest
dreams of the most morbid fiction writer, and will ever live
as perpetual reminder to the thoughtless of the uncertainty
of existence. Owing to the social prominence of its
one hundred and fifty victims, this catastrophe
stands out unique in the annals of great fires. Imagine the elite of a great city, ·the subscribers to
such fashionable organizations as the Opera, the
Horse Show and, in England, Ascot. Pack them all within
a limited area, apply a match and make a bonfire of the
surroundings and picture the result. These formed the
patrons at the bazaar in question, when, at 4 P.M. on that
day, hundreds of persons were crowding the narrow aisles
between the stalls decorated to represent the streets of old
Paris, and were gazing with interest at the many titled men
and women who had offered their services on behalf of a
well-deserving charity. The building itself was a one-story
wooden structure with a freshly tarred roof, and contained
draperies and curtains of highly inflammable material. As
in most of these instances, the origin of the fire is doubtful;
it may have been caused by the overturning of a spirit lamp
or the ignition of the illuminating apparatus of a cinematograph which had been installed for the additional amusement of the visitors, but all that is definitely known is, that
at this hour in the afternoon an explosion took place on the
left side of the bazaar. The flames, seizing the hangings

and articles exposed for sale, spread rapidly, and the crowd instinctively sought the farthest point from danger. Of the eight doors, one was on the left, and therefore cut off by the flames, three opened on to the Rue Jean Goujon and four, located in the rear and used by employees, were unknown to the guests. People near the main entrances were able to escape with but slight injury, but the great mass of humanity surged towards the right wall where there was no outlet save a small window heavily barred, which connected with the Hôtel du Palais. Servants in the hotel, who had been peering through this opening to obtain a glimpse of the gay throng, succeeded in breaking the bars and rescuing a number of the panic-stricken throng, but while so doing many were burnt before their eyes.

The first intimation of the situation to passers-by was a rush of semi-nude and maddened women into the adjacent streets where instantly all became confusion. Rows of stately carriages and humble cabs, whose drivers had been awaiting the arrival of their employers, were roused into activity by the vision of their shrieking, blood-stained owners, wildly clamouring to be driven anywhere away from the scene of horror. Grooms in the service of the Baron de Rothschild, whose stables were nearby, used their hose to good purpose in extinguishing the flames enveloping the filmy gowns of escaping patrons, and one man, more clear-headed than the rest, plunged at full length into a horse trough to find relief from his sufferings. Before the firemen could arrive the whole structure was in a blaze and the building collapsed even as the engines galloped up. It had been known to the authorities that the hall was anything but fire-resistant, though being built upon private property they had not been able to take any steps in the matter, and it had been thought that its dimensions, and the fact that it was on the street level, was sufficient guarantee of its security.

In the meantime rescues had been effected in the interior by a few brave priests, who, by means of some ladders, had led about thirty persons over the walls of a neighbouring convent. But anything in the nature of organized fire-fighting was out of the question, the flames having got beyond control, and the whole structure resembling nothing so much as a giant funeral pyre, which was intensified by the piteous moans and cries for help, which no human power could give. It is difficult to gather any collected narrative of what happened within. In moments such as these impressions are fleeting and as elusive as the phantasmagoria of delirium. But a few episodes remain, illustrative to some extent of the nature of the struggle for life, while others exemplify the height of self-abnegation, to which on occasion individuals arise. The story of the martyrdom of the Duchesse D'Alençon was related afterwards by an eye-witness, a young girl who had been assisting her at a stall not far from the outbreak of the fire. As the younger woman saw the flames approach she begged her friend to escape, pointing out the fact that the main entrance was near, and that the fire would soon be upon them. But the Duchess replied in calm tones that it was their duty to allow the visitors a first chance, and she and her terrified companion remained at their post watching the waves of frightened people beat their way to safety, until the heat became so intense that Mademoiselle L. could endure it no longer. With one last entreaty to the Duchess she joined the others leaving her brave companion with hands clasped across her breast and eyes steadfastly fixed on her approaching doom never to be seen again alive. It may here be remarked that the Duchess was a sister of the Empress of Austria, who later was to die a victim to the assassin's knife, and that both were universally known and beloved. Some may find food for reflection in the extraordinary manner tragedy appears to dog the footsteps of the members of certain families, and of a truth fire is no respecter of persons, as has been instanced again and again. When the firemen were able to enter the ruins of this charnel house they found near the fatal right wall a mound of dead five feet in height, denuded of clothing and many unrecognizable. The Duchess was identified only by a ring and certain stopped teeth in her jaw. Piteous was the plight of many of the survivors, some of whom became insane from fright, while others were so severely injured that they afterwards died or carried traces of the experience for many years.

It is out of the question to criticize what might, or might not, have been done in the case of a disaster of this nature. With a non-fire-resistive structure and conditions such as prevailed, from the first the case was practically hopeless, though, as a counsel of perfection, had panic been avoided more persons might have been saved, and notices advising visitors of the back exits should have been displayed. But even the latter would probably have availed little, since it is the prime impulse of every person in a building to leave by the exit through which he or she entered. This it is which makes it of supreme importance to have properly drilled aisle guards and a staff who, in emergency, will keep cool and act as pilots to the excited and hysterical. It is not too much to say that if all were possessed of the splendid courage of the Duchesse D'Alençon, less life would be sacrificed to fire.

It is a relief to turn from the contemplation of such horrors to a conflagration, which, if involving tremendous financial loss, at least was unattended with the harrowing scenes which have been described above. In London on the 19th of November, 1897, a fire broke out at 30 Hansel Street, in the heart of the manufacturing and warehouse section of the city. The origin of the conflagration was the explosion of a gas engine on the premises of a large firm of mantle manufacturers. The employees, terrified by the smoke, rushed to the roof and fled shrieking in fear over the adjoining buildings. A strong wind was blowing, and, as is often the case in emergencies of this nature, everybody's business being nobody's business, there was some delay in transmitting a fire call. On the arrival of the Brigade the flames had spread to a neighbouring warehouse and had crossed the street to a paper factory. In this part of the city the streets are particularly narrow and great difficulty was experienced by the firemen in conveniently placing their apparatus. Large forces of police were re-

quired to keep back the crowds who sprang up as if by
magic and threatened seriously to hamper the operations
of the fire-fighters. One after another the buildings, stocked
with large supplies of novelties and goods for the Christmas
market, were involved, and an explosion of gas meters
added to the complexities of the situation. Firemen, who
had ascended to the roofs of fire-free buildings in order
better to attack the outbreak, found their retreat cut off,
and the excited spectators witnessed many daring rescues
of these brave men by their comrades. The vicarage of
St. Giles Church, Cripplegate, was completely destroyed,
and the church itself, interesting on account of its historic
associations, was saved after almost superhuman effort. In
all, one hundred houses covering four acres were consumed,
and the combined exertions of practically the entire brigade
were unsuccessful in checking the flames until 5.30 P. M.,
when a wall collapsed in Well Street, arresting the progress
of the latter. The width of Red Cross Street was fortunately a sufficient barrier at that point, for had the fire
broken through it is impossible to say where and how it
would have been stopped.

Some idea of the magnitude of the conflagration can
be gleaned from the fact that at midnight no less than fifty
engines were still at work, and the fire was not under complete control till the following morning. The total financial
loss amounted to five million pounds ($25,000,000), put two
thousand people out of work and sent up the price of
ostrich feathers in all parts of the world. There is an
absence of spectacular detail about such an outbreak, which
tends to make it almost dull and uninteresting, but at the
same time it illustrates effectively the vast risks which are
to be found in European towns and goes to show that the
London Fire Department, though to American ideas lightly
equipped as regards personnel and apparatus, is at times
called upon to fight fires of the first magnitude. It is perhaps this very absence of spectacular effect which makes
the realization of fire peril so difficult to the European
and so vivid to the American. Baltimore, San Francisco,

DANGEROUS WORK.   NOTE CRUMBLING WALL.

the Equitable, were occurrences of world wide interest and absorbed the descriptive talent of every skilful writer on two continents. A fire such as the above is merely a record of good work well and bravely done in the most unromantic of surroundings, and with a total absence of "colour," pathetic, exciting, or enthralling. The business of the world was not temporarily dislocated, though the pecuniary values involved were so tremendous; lives were in danger, certainly, but so they are daily and the fact passes unnoticed. Hence it is that in describing great conflagrations, those in Europe are apt to sink into insignificance and those in the States loom out large in their gaunt and staring hideousness.

In this respect it may not be inappropriate to add a few words about the fire danger in conjunction with floods. In the spring of most years and alas! particularly in that of 1913, floods often occur through the rising of rivers and vast tracts of territory are inundated, while towns and cities are washed away or destroyed by fire. That latter phrase often gives rise to comment. People argue, how can it be possible to have fires when it is water which is giving the main cause for alarm? The answer is simple enough: gas mains burst, oil stoves are upset, electric light mains are severed and become potential torches and there is no means of effectively fighting the outbreak. Streets, impassable through water, naturally prevent the operating of any but floating fire apparatus, and thus it is that flames and flood sometimes work as allies, and humanity stands staggered at the immensity of the forces combined against it. But there is one comforting reflection, that silver lining which borders every cloud, namely that year by year the services of science are being called upon to a greater degree to keep within control the latent forces of nature. Houses are built fire-resistant, apparatus is perfected, waters are dammed, rivers are banked, and inch by inch, day by day, the never ceasing combat continues till the time shall come when the victory shall lie with man. That day will dawn, of that there is no doubt, and the swiftness of its advent will be exactly proportionate to the determination of the human race.

Amongst some of the great conflagrations known to history, the following are representative, though it may be hazarded that the financial values involved must, in the earlier years, have been problematical, as when an entire city is wiped off the map, it is obviously difficult to total even approximately the fire loss. Ancient Rome boasts of one great outbreak which consumed almost every building within its walls, this in 64 A. D. Constantinople might not inaptly be described as the much burned, since it had three conflagrations in the eighteenth century alone, one costing one hundred lives and fifteen thousand dwellings, another three hundred lives and $30,000,000 worth of damage, and the third, thirty thousand dwellings and a property loss of $115,000,000. Moscow, outside of 1812, when the city was destroyed by its own inhabitants, rather than allow it to fall into the hands of Napoleon, was wiped out in 1383, the destruction on this occasion being even greater than the later event, since naturally the construction was inferior. The great fire of London occasioned a property loss of $60,000,000, while in 1861 the business section suffered to the extent of $12,000,000, and in 1874 the residential area suffered to the extent of $15,000,000. A conflagration of gigantic proportions gutted Smyrna in 1796 and destroyed half the city with a loss of over $50,000,000. Turning to America. The great fire of New York in 1835 destroyed six hundred buildings with a loss of $20,000,000, while that of Boston in 1872 represented the second highest total extant, namely $100,000,000. The record for fire loss, before the conflagration in San Francisco, was held by Chicago, which in 1871 lost 17,500 buildings and $200,000,000 worth of property, though without appreciable loss of life. Toronto in 1904, St. John's in 1892, and Hamburg in 1842, were also visited by serious outbreaks, that in the German city burning all the business section with a loss of $35,000,000, while the Newfoundland capital suffered to the extent of $26,000,000, a remarkable figure, taking into consideration the small size of the town and the relatively minor importance of its financial values. After such a recitation, who shall say that personally, financially or structurally, fire does not constitute one of the greatest perils extant?

CHAPTER XII

THE HOTEL PERIL

WITHIN the last twenty years a great change has come over family life, both in Europe and America, and the reign of the hotel seems established. Everywhere vast caravanserais are springing up, and though replete with all the comfort the mind of man can devise, and though advertised as "fireproof," their construction is often such as to render them an easy prey to fire and therefore dangerous to human life. That some people are aware of this fact is evidenced by the frequent demand of visitors for rooms "not too high up," or "on the lowest story." For, it must be remembered, that people do not perish only by fire itself, but from suffocation consequent on smoke, from ill-judged action caused by panic and from other indirect causes. Also, the expression "fireproof," as applied to a building, does not include its furnishings and equipment, and is further no guarantee that it has been designed along the lines of greatest resistance to the fire peril. Finally, the fireproofing of materials is not always satisfactory, and a story is told of a contractor in that business who was asked by a friend what was done with all the shavings and chips from fireproof wood. The nonchalant reply, "We use them to light the stoves in the morning, they make excellent kindling," gave him food for reflection. There can be no doubt that hotel fires are extremely prevalent, as may be judged from the following figures. During the first day of 1913, five hotels in widely separated portions of the United States were destroyed, with a loss of two lives and $100,000. The total of such fires in the month of January was twenty-five, representing a property loss of $700,000, and seven lives. In 1912 there was a hotel fire every thirty-three hours in North America, and up to date, 1913, that record has been passed, with an outbreak every thirty hours. It has been estimated that the property loss in the United States, through these disasters, during the last five years has amounted to $25,000,000 (£5,000,000), while the death roll has been proportionate. These figures, it is true, apply to America, but similar occurrences are common enough in Europe, and are by no means confined to the older-fashioned structures. To wit, the fire at the Carlton Hotel is still fresh in the memory of Londoners.

Now it must not be supposed that this state of affairs is due to the apathy of hotel proprietors and managers as to the safety of their clients; apart from considerations of humanity and sentiment, that would be bad business. Rather is ignorance the root of the evil, ignorance of the very first principles of fire control, which all responsible for the lives and safety of others should thoroughly understand. It is only too common to find an attic at the top of an hotel used as a lumber room and filled with all kinds of inflammable rubbish such as old mattresses, empty boxes, excelsior and waste paper, a perfect magazine of combustible material and a direct invitation to a visit from the flames. Many hotels again, have unprotected elevator shafts around which circle the main stairs; should a fire originate on the ground floor, instanter the shaft becomes a flue up which the flames sweep with amazing rapidity, and the stairway as a means of exit becomes impassable. Defective electric wiring is likewise a constant source of danger, short-circuiting constituting one of the most serious of risks. As for heating apparatus, with faulty connections, improperly covered or wrongly situated hot-air ducts— were this cause of trouble eliminated, it is no exaggeration to say that hotel fires would decrease by one-third. It may be imagined that the introduction of precautions necessary to combat this peril spells the expenditure of large sums of money and radical structural alterations. Broadly speaking, this is not the case, the expenditure of a certain amount of common sense and care will produce far reaching results, as the history of hotel fires goes to show, while in the case of new construction it should be realized that "skimping" in the matter of fire protection, in the long run, is the worst kind of "penny wise, pound foolish" policy. The municipal authorities, of course, insist upon compliance with certain regulations when the erection of a hotel is undertaken, varying with the country and local conditions, but as a rule the building code is directed chiefly towards insuring safety of exit for guests, rather than interfering with the larger issue of how the necessity of a hurried exit may be avoided. At the same time the writer must place on record the fact that, in New York, the new hotels embody every known means of fire resistance and are as perfect in their construction as the present state of human knowledge will allow.

But what precautions then should be taken in older buildings and are they beyond the scope of the average manager? The answer may be framed in the form of another question or rather series of questions. Has everything been done to prevent a possible outbreak by the removal of potential sources of the same? This is largely a matter of common sense coupled with some thought. Then, can a fire be readily detected? Is there an automatic fire alarm or is there a night watchman who records his tours of inspection in a clock? Can guests be readily alarmed, and is there direct telephonic communication with the fire department? Is there an efficient system of fire escapes and is the house properly provided with chemical extinguishers and such like apparatus? Are the employees competent to deal with an incipient blaze and have any regulations been issued as to the particular duty of each in the event of an emergency? These suggestions do not represent a considerable capital outlay and yet are all of primal importance. Of course, it is easy to continue the catechism further and to ask whether, in design and construction, the building is such that it is feasible to confine a fire within certain limits, whether elevator shafts are covered in, whether floor openings are unprotected, whether there is a sufficient water supply, and whether the house is guarded against exposure fires, i.e., fires caused by adjacency to some other burning structure, a common enough contingency and one easily met by the adoption of wire-glass in windows? This may appear a formidable battery of queries, but a little consideration will suffice to show that their bark is worse than their bite, and that, after all, there is nothing so dreadfully radical in the proposition as to necessitate loss of sleep or visions of speedy bankruptcy. The great conflagrations of the world have not been due to elemental disruptions, as a rule, beyond the control of man, but rather to acts of deliberate carelessness or thoughtlessness, which might easily have been avoided. And so it is with fires in hotels; they constitute a real peril which annually reaps a rich harvest of lives and property,—a minimum of precaution and the harvest would not be garnered.

The following examples of hotel fires which might have been avoided are selected from a list prepared by "Insurance Engineering," a monthly publication devoted to the science of fire control. "Brockville, Ontario, Canada. Strathcona Hotel. Cause, overheated furnace in basement. Discovered by clerk at 4.45 in the morning. No private appliances. Fire Department handicapped by delayed alarm and lack of sufficient apparatus with which to fight the fire. Loss considerable." Overheated furnaces are a source of such constant trouble that the heating plant should always be isolated and situated in a fireproof room, though a case is recorded from Chicago in which it was found that the heat from the firebox of a boiler was so intense that it ignited some sheets of music on the other side of a thick brick wall. Hence, isolation cannot be too carefully ensured. "Chicago. York Hotel. Cause of fire, defective electric wiring. Discovered by watchman, 3.16 A. M., in partition in first story. Fire department immediately notified. Fire spread to roof in hollow finish. Private fire protection poor. Firemen who arrived promptly helped guests to escape by stairs. Loss nominal, owing to the prompt and effective work of the fire department." Defective electric wiring is too frequent a cause of fire, and can easily be avoided by regular inspection. It is then the safest method of illumination in the world. A word may here be inserted about "hollow finish." This is the system whereby spaces are left between the outside covering of a wall, ceiling or floor in the main constructional work. Such cavities, if subjected to fire, are a source of serious danger, since the air therein encourages the flames, whereas, if built up flush, this danger disappears. "Rimouski, P. Q. St. Germains Hotel. Three story, wood. Cause, hot stovepipe on the floor of the second story. Fire spread through hollow wall finish. Loss, total." "Charleston, Ont. Grand View Hotel. Cause, oil heater in pool-room. Fire spread to other buildings and caused a conflagration. Loss, $200,000." These are good examples of how fires occur through defective heating arrangements. It seems scarcely necessary to insist that in any building, stovepipes should be most carefully protected, while oil stoves, as heaters, should be abolished *in toto*. "Akron, Ohio. Thuma Hotel. Five stories, brick, ordinary construction, hollow finish. Cause, grease fire on range of kitchen in basement ignited coating of grease in vent shaft which passed upward through building, part of the way between the ceiling finish of the second story and the floor of the third. Fire Department responded quickly to a box alarm and fought fire for six hours. When the firemen arrived the fire was general throughout the building. Owing to the effective work of the firemen the loss was limited to twenty-five per cent. of the values." Vent ducts from kitchen ranges are peculiarly liable to ignition since, in course of time, the pipes become coated with a thick deposit of inflammable grease. Should this catch fire, great heat is generated, and the duct becoming red hot, will ignite any wood adjacent to it. Hence, every precaution should be adopted for the isolation of these vents, so that, in the event of an outbreak, they may burn out without causing more serious trouble.

Of the inconsequent carelessness of hotel employees, a whole volume might easily be compiled. The following are, however, good examples: "Salina, Kansas. National Hotel. Fire started in the basement, in laundry chute into which a cigar butt had been thrown. The chute was of wood and extended from basement to roof, with unprotected openings in each story. The fire was discovered by the hotel porter, but an alarm was not sent to the fire department. The notification to which it responded was the fire itself which was seen by several firemen. The hotel had been inspected by the fire department, and the owner warned against the dangerous construction and arrangement of the chute." "Missoula, Montana. Florence Hotel. Three story, brick, ordinary construction, hollow finish, unprotected floor openings. Fire started in elevator shaft in the rear of building, and was caused by a can of hot ashes set on the platform of the elevator car. Fire was discovered at 11 A. M. by a clerk, who promptly transmitted the alarm to the fire department. The flames traveled up elevator shaft, and 'mushroomed' in the attic, between the ceiling of the top story and the roof. A partition in the attic, between the main building and a wing, assisted the firemen in checking further spread of fire. It took five hours to suppress the blaze."

The carelessness of hotel servants is proverbial, and to make them realize the danger of the thoughtless throwing away of an oily rag, the improper disposal of rubbish, or of an unextinguished cigarette or cigar end may not inaptly be compared with the labours of Sisyphus. When it is remembered, that in some large hotels the staff employed number about two thousand souls, the extent of the mischief can be gauged. And if servants are careless, what of guests? Contemplate the following: "Tacoma, Washington. Grand Hotel. Four story, brick, ordinary construction. Fire started at 5.35 P. M., and was caused by a man smoking in bed. It was discovered quickly by other guests, and the fire department responding promptly, controlled the outbreak so that the loss was limited to $17,000." Comment really seems to be needless and the protection of the individual against himself has not added to the lightening of the burden of those responsible. But probably the most terrible exemplification of the mischief which can be wrought by a thoughtless visitor is embodied in the story of the Windsor Hotel fire. This building occupied the entire block on the east side of Fifth Avenue in New York City, between Forty-sixth and Forty-seventh streets. It was of antique construction, with wide halls, high ceilings and several elevator shafts. On the 17th of March at 3 P. M. a guest in a front parlor on the second floor lighted a cigar and threw the still blazing match into the street. As it passed the curtains, the latter ignited and in an instant were in flames. Without attempting to extinguish the blaze or to give an alarm, the author of the disaster fled from the room, and a few moments afterwards the head waiter, in passing the door, caught sight of the fire, which, by that time, had greatly increased. Unaided he made a brave effort to subdue it, but his hands were badly burned and it was easy to see that more help was needed. The St. Patrick's Day parade was passing at the time. The streets were lined with spectators and guarded by policemen, interested onlookers were leaning out of the windows of the hotel itself and the strains of many brass bands deadened all other sound. As the head waiter calling, "Fire," ran into the street and endeavoured to reach an alarm box which, unfortunately, was situated on the other side of Fifth Avenue, he was prevented from crossing by a puzzled policeman, who could not understand the excited man's incoherent explanations above the din of the music. But the smoke and flames soon told their own story and a first, second, and finally a fourth alarm were sent in. Owing to the construction of the building the flames ascended both by way of the halls and in and out of windows to the top floor with great rapidity. In spite of the desperate efforts on the part of the fire department, who were handicapped by a poor water supply, before 4 P. M. the hotel was in ruins. A little later the only wall to remain standing slid down to its base like a closing fan. By 7 P. M. the fire was under control and the safety of adjoining property was assured. Of the many guests and servants who had been watching the procession, fourteen were dead and about fifty injured. Some of them had attempted to use the safety ropes which had been placed in each bed-room, but the friction on their hands became too great and they were forced to let go and meet their doom in the streets. One handsomely dressed woman on the fourth floor held out her arms as though imploring aid from above, then without a cry she jumped, turning over and over as she fell until she struck the iron railing below. At one window appeared a woman bearing in her arms a child. Terrified by the flames which were licking the sill from the floor beneath, she threw the child into the street and an instant later followed. Many rescues were effected by the firemen who mounted on ladders and dragged to safety some of the occupants, and if others had not been panic-stricken by the proximity of danger, and had possessed sufficient courage to await the arrival of help, many of those who jumped to death might have been saved. Behind the hotel, and connected with it from the interior, was a Russian bath establishment, where a number of patrons were enjoying the pleasures of treatment. They were obliged to make the best of their way out clad in sheets, towels, or whatever articles of clothing were nearest to hand. Two men in the hotel who were vainly hunting for a fire escape were met by a trained nurse, who said that she could conduct them through her room to the object of their search. When they had entered, however, she put her back against the door and told them that they must assist her in carrying her patient, a helpless old lady in a wheeled chair, to a place of safety. In other words, this plucky woman had invented this scheme in order to save the life of her charge, and the men, infected by her courage did as requested, and all four gained the street without mishap. All this owing to an act of carelessness on the part of a visitor, whose identity, by the way, has never been discovered to this day.

Prevention is, of course, better than cure, but next to that is promptness of action, both direct and indirect. That is to say, an outbreak of fire should be detected as soon as possible, which may be accomplished either automatically by sprinklers, by a watchman, who registers his inspection visits on a clock, or by both. It must never be forgotten that every minute lost means ten times the additional risk. The following type of case is, unfortunately, too common. "Sioux City, Iowa. Mondamin Hotel. Four stories brick, ordinary construction, unprotected floor openings. Fire · started 8.20 P. M. in boiler room in basement. Discovered by outsider who transmitted alarm. Since discovery of the fire was delayed, fire department was unable to control it. Loss, $120,000." A watchman, at ten dollars a week, would not have been an extravagant rate of insurance.

Again, contrast the following: "Lansing, Michigan. Downey Hotel. Six story, brick, ordinary construction, hollow finish, unprotected floor openings. Cause of fire, a heated bearing in, or an electrical defect at, elevator motor in pent house over roof of elevator shaft. Discovered, 5.59 P. M. by hotel employee. Alarm received by fire department, 6.25. Fire burned until 8 A. M. next day. Loss over $100,000." "Little Rock, Arkansas. Gleason's Hotel. Four story, brick, ordinary construction, hollow finish, unprotected floor openings. Fire caused by electric motor top of elevator shaft. Discovered by employee at 1.08 A. M., box alarm transmitted immediately. Fire controlled in thirty minutes, and confined to locality of origin. Loss, $2,300, less than three per cent. of values." A better exemplification of the advantages of prompt action could not be imagined. The notifying of guests in hotels of an outbreak of fire is of supreme importance, since, as a rule, such outbreaks occur at night when most of the inmates are asleep. It is a good scheme to have an alarm gong fitted in the bed-rooms, which should be operated from the reception bureau or other central position. But even such methods should be supplemented by personal calls from members of the staff. This will go a long way towards

QUICK BURNER.

preventing a panic, of which there is a danger if the gong alone is used. As for fire escapes, this is a vast and intricate subject. Time and again have persons been injured on narrow fire escapes, while, as stated in the Windsor hotel fire, a rope provides only a last and desperate means of exit. Some hotels are now erected with fire escape towers, which completely cut off the flames and insure an open road to safety. But it is impossible to lay down any hard and fast rules for the construction and placing of contrivances, since, to a certain extent, the design of the building must be taken into consideration. And in all cases sufficient and careful thought should be given to these matters. It seems hardly credible that there should be hotels devoid of even a hand chemical grenade, yet fire chiefs frequently report that such is the case. Every establishment of a certain size should not only be properly equipped with hand and chemical extinguishers, but should also be possessed of a private fire department. The formation of such an organization offers no particular difficulty, and in the opinion of the writer, is as worthy of advertisement in hotel announcements as such hackneyed phrases as "unsurpassed cuisine," "moderate terms," and "unrivalled view." The casual visitor would sleep just as soundly were he deprived of those three remarkable benefits, but he might be forgiven for passing a restless night were he haunted by the terrors of fire due to "poor fire control." And now to come to an all-engrossing portion of the theme under discussion, namely, why fires spread rapidly in hotels. In nearly all such buildings there is a lack of subdivision of floor area, though in some cities an interior wall of incombustible material is required between every set of four rooms, this extending from foundations to roof. In one of the latest New York hotels, the partitions between rooms are of hollow tile, the doors of steel and the transoms glazed with wire-glass; even the trim and picture mouldings are of metal. That this is the very height of perfection in fire control may be gathered from the fact that in this same hotel an outbreak recently occurred amongst some furniture stored on an upper floor. The furniture was completely destroyed, but the room was habitable twenty-four hours later, while the adjoining premises were unharmed. Unprotected floor openings, like the poor, are ever with us, and embody the most glaring structural defects imaginable. Their retention is virtually a crime, especially considering the facility with which this risk can be remedied. Cases without number might be cited of the prominent part played by this avoidable fault in hotel conflagrations, but the two following may be quoted as typical. At a hotel in a Kansas city, the stairway encircled the elevator shaft—a form of suicidal internal architecture peculiarly popular in England and on the continent of Europe. The fire started early in the morning in the basement, cutting off the escape of the guests, many of whom jumped from windows, while others slid down ropes made of bed clothing. The other hails from Oneonta, New York. "Central Hotel. Fire discovered at 3.30 A. M. under basement stairs by clerk. ▪ No private fire appliances. Fire department handicapped by wires in street. Rope fire escapes only. Three lives lost." In such terse language is summed up the result of unprotected floor openings. Fire and smoke naturally ascend, and hence it is of paramount importance that not only should stairways and elevator shafts, dumb waiters, pipe and wire chases, be of fireproof construction, but each opening should be entirely enclosed by fireproof materials. Elevators and stairways should always be separated; the encircling stair and the lattice work elevator shaft being an invention of the fire fiend himself. The shaft of an elevator may well be compared to a factory chimney. Every one knows that the giant smoke stacks which dot the hillsides of any manufacturing neighbourhood, have not been erected with a view to the picturesque. Rather is their purpose strictly utilitarian, the higher the chimney the greater the draught, the fiercer the fire and the more tremendous the heat. It is exactly the same with an elevator shaft with a fire at the bottom, which, if closed at the top has the effect of drawing up the smoke and heat, which form the primal obstacles to escape by inmates on the upper floors. Thereafter, the fire spreads laterally and downward. Hence, these shafts should be rendered as completely "fire tight" as compartments in a ship are constructed "water tight."

Finally, elevator machinery should be placed at the top of a shaft, as the lubricating oil and grease used on its running parts form ready material for the flames. The same may be said to apply to stairways, though in this connection it may be remarked that particular attention should be paid to the basement and attic entrances of the same, as it sometimes occurs that these are left unguarded, and these two points constitute, as a rule, the beginning and the end of hotel fires. Interior light courts are also a source of danger, especially when roofed over. All windows looking onto such courts should be glazed with wire glass, and, as a matter of fact, light wells should never be roofed. As regards "hollow finish," the following two examples explain the danger more succinctly than columns of technicalities: "Putnam, Connecticut. Chickering Hotel. Three stories and basement, wood walls, ordinary construction, hollow finish, mansard roof. Fire started in basement near boiler, discovered at 1.30 A. M. by a passer-by. Burned six hours. Loss $19,000, about 55 per cent. of values. Chief of Fire Department said: 'The fire worked up inside partitions to the roof. There was not a square yard of flooring burned in any place.' " "Excelsior Springs, Missouri. New Elms Hotel. Three story and basement stone building. Fire started at 1.30 A. M. in coal bin outside of building. Discovered promptly and quick alarm sent in. No private fire protection and fire department handicapped by weak water pressure. Fire Chief's reasons for spread of fire as follows: 'There were no fire walls in the building. There were wide spaces between ceilings and floors to act as "deadeners," and it was through these spaces that the fire spread through the building and made it difficult for firemen to get water at the right place at the right time.' " This system of introducing "deadeners" is a concession to the visitors, who naturally enough dislike noise, and who otherwise would be disturbed by their neighbours. It can be rendered safe or at any rate partially so, by filling up these spaces either with asbestos or mineral wool. Ventilation systems should also be carefully supervised, as on occasion they may prove responsible for serious fire risk. The following instance is illustrative of the care which must be exercised over hotel design, where, be it remembered, panic is above all else to be avoided. In a New York hotel a huge volume of smoke suddenly filled a crowded dining room. The cause was the burning of a heap of rubbish which had been placed too close to the air intake of the ventilating fan, which draughted the smoke and blew it on through the ventilating system. Nothing more serious than the annoyance and discomfort of the guests resulted, but the draperies and decorations were damaged by smoke. Had the intake been located higher up, or had it been arranged to close with movable louvers the trouble would not have occurred. Fire exposure or the danger to be apprehended from fires originating nearby and in turn communicating with an hotel can to a great degree be guarded against by the fitting of window openings with hollow metal sashes glazed with wire-glass. That this risk is not so remote as might be supposed may be seen from the following: "Oakland, California. St. Mark's Hotel. Eight stories, reinforced concrete. Fire started in sign painter's shop in second story of adjoining building and burned out windows of hotel, which were sashed with wood." "Kansas City, Missouri. Ormond Hotel. Five stories, brick. Fire originated in garage adjoining, between ceiling of first and floor of second story. Cause, defective electric wiring. Garage employees delayed sending in an alarm. Fire department handicapped by headway of fire, height of hotel and weak water pressure. Insurance loss, $140,000; values $310,000." It goes without saying that hotels as frequently burn other buildings and that these remarks may be taken as being applicable to all houses of whatever type. Of course, it may be urged that this use of wire-glass is deplorable from an æsthetic point of view, which with some people counts for more than common sense and the protection of life and limb. For such artistic souls it is impossible to cater, though it is fortunate that with the majority of the community fire risks are more important than landscapes, however inspiring.

Which introduces the conclusion of the subject. It has been demonstrated *ad nauseam* in the preceding pages that hotel fires are very real contingencies against which to prepare, and it has been shown that the fireproof hotel is not yet to be considered as practical politics. But it can be made fire resistive and that with a degree of certainty which will minimize the risk to an appreciable extent. The automatic sprinkler will do everything except start a fire. As explained elsewhere its construction is simplicity itself, while not only does it automatically damp down an incipient blaze, but in addition will operate a fire alarm, insuring that there is no delay on the part of either employees or fire department in tackling the enemy. It is perfectly possible to install this system in the public rooms of an hotel and yet interfere not at all with the decorative scheme, which would be treason in the eyes of some. In one building so protected, the sprinklers number no less than 1,600, the source of water supply being a 20,000 gallon tank elevated twenty-five feet above the roof, and two six-inch connections with the city main. By this method it is possible for a room to be burnt out and the fire subdued, without the damage to property and the excitement amongst guests which would be caused by the arrival of a brigade and the subsequent operating of hose pipes through the hall and stairways and through windows. The sprinkler system is, in fact, the silent guardian of life and property, which "slumbers not nor sleeps" and which can be relied upon as a rule. A rise in temperature 160 degrees Fahrenheit on the floor is sufficient, and the sprinkler starts to work, sending down a drenching stream upon the affected area and warning all and sundry that there is an enemy at hand. At a recent fire in a hotel guarded in this fashion, one of the guests rang and complained of a water pipe located just above his bed, which had burst suddenly and awakened him from his beauty sleep. His indignation was unbounded and in the morning he demanded an apology from the manager, which was smilingly forthcoming. But that individual did not think it necessary to explain to the irate guest that the room above his (an unoccupied one) had caught fire and that the lives of some five hundred guests had been quietly and quickly saved by an inconspicuous "sprinkler."

CHAPTER XIII

THEATRES AND FIRE PANICS

THE problems affecting 'fire control in places of public amusement are amongst the most intricate demanding solution by Fire Departments. For here the human element becomes an important factor in the situation, and though every safeguard scientific ingenuity can devise may be adopted, and though thousands of dollars may be expended in the installation of the most modern and complete equipment of that nature, it lies within the power of one small boy in the gallery, who thoughtlessly calls out "Fire," to transform an assembly of happy pleasure-seekers into a shambles. That is to say, unless some scheme of controlling an audience in moments of emergency can be devised, and towards this end many fire protection associations are working. Hence in treating the subject it will be convenient first to consider the active measures demanded by municipalities from the managers of theatres for the public safety, then to give an example of an actual theatre fire, drawing from it the obvious deductions, and finally to touch on audiences themselves.

Broadly speaking, theatre safety depends upon the situa tion and convenience of exits, the use of the fireproof curtain completely separating the stage from the auditorium, the installation of a system of automatic sprinklers in places where much inflammable material is to be found, such as in scenery docks, and minute attention to such details as the provision of fireproof scenery, and the caging in of all lights, electric or otherwise. Perhaps it will be simplest to give the regulations suggested for or existing in

New York on this subject, which may be accepted as representing the standard requirements.

Standpipes four inches in diameter must be provided with hose attachments on every floor and gallery; one on each side of the auditorium in every tier and one on each side of the stage in every tier. In addition there must be at least one in the property room and one in the carpenter's shop, if the latter be contiguous to the building. All these standpipes must be kept clear from obstruction and be fitted with the regulation couplings of the fire department. They must be kept constantly filled with water by means of an automatic power pump of sufficient capacity to supply all the lines of hose when operated simultaneously. This pump must be ready for immediate use at all times during a performance. A separate and distinct system of automatic sprinklers with fusible plugs, supplied with water from a tank situated on the roof over the stage and not connected in any manner with the standpipes, must be placed on each side of the proscenium opening and on the ceiling over the stage at such intervals as will protect every square foot of stage surface when they are in operation. Wherever practicable these sprinklers must also be placed in the dressing rooms, under the stage and in the carpenter's shop, paint rooms, store and property rooms. A sufficient quantity of hose fitted with regulation couplings and with nozzles and hose spanners must be kept attached to holders. For immediate use on the stage there must always be kept in readiness four casks full of water and two buckets to each cask; all to be painted red. There must also be provided hand-pumps or other portable fire extinguishing apparatus, and at least four axes and six hooks of different lengths on each floor of the stage.

Every portion of the building devoted to the accommodation of the public, as also all outlets leading to the streets, must be well and properly lighted during the performance, and the lights must not be extinguished until the entire audience has left the premises. The illumination

MODERN THEATRE.   NEW YORK CITY.

of all parts of the building used by the audience, with the exception of the auditorium, must be controlled from the lobby by a separate shut-off. Gas and electric mains supplying the theatre must have independent connections for the auditorium and the stage, and provision must be made for shutting off the gas from the outside of the building. All suspended or bracket lights surrounded by glass in any portion of the theatre used by the public must be provided with wire netting protection. No gas or electric lights must be inserted in the walls, woodwork, ceilings or in any part of the building unless protected by fireproof materials. The footlights, when not electric, in addition to the wire network, must be protected with a strong wire guard and chain placed not less than two feet distant, and the trough containing the footlights must be composed of and surrounded by fireproof material. All border lights must be subject to the approval of the Department of Buildings and be suspended for ten feet by wire rope. All stage lights must have strong metal wire guards not less than eight inches in diameter so constructed that any material in contact therewith is out of reach of the flames. The bridge calcium lights at the sides of the proscenium shall be enclosed in front and on the side by galvanized iron, so that no drop can come in contact with the lights. Electric calciums, so called, are included in the above requirements. Standpipes and all apparatus for the extinction of fire or for guarding against the same, must be in charge and under the control of the fire department, and the Commissioner is responsible for the carrying out of these regulations.

A diagram of each tier, gallery or floor showing distinctly the exits therefrom, each plan occupying a space of not less than fifteen square inches, must be legibly printed in black lines on the programme of every performance. Every exit must have over the inside of the door the word "EXIT," painted in legible letters not less than eight inches high. All exit doors must open outwards and be fastened with movable bolts, which must be kept drawn during performances. No doors shall open immediately upon a flight of stairs, but a landing of reasonable width shall be allowed between them. The proscenium opening must be provided with a fireproof metal curtain or one constructed of asbestos, overlapping the brick proscenium wall at each side not less than twelve inches and sliding vertically at each side within iron channels of a depth of not less than twelve inches. These grooves must be securely bolted to the wall and must extend to a height of not less than three feet above the top of the curtain when raised to its full limit. This curtain should be raised at the commencement of each performance, lowered between each act and again lowered at the end of the performance. This system is now regularly in force in Chicago. If the curtain be made of asbestos, that material must be reinforced with wire, while to ensure its remaining taut and its easy descent, a rigid metallic bar of sufficient weight must be firmly attached to its base. The excess weight of the curtain is to be overcome by a check rope of cotton or hemp, extending to the floor on both sides of the stage, so that its cutting or burning will release the curtain, which will then descend at its normal rate of speed. This curtain shall at no point be nearer the footlights than three feet.

As regards doorways, none shall be allowed through the proscenium from the auditorium above the first floor and all doorways shall have self closing standard fire-doors on each side of the wall. Openings, if any, below the stage must each have self-closing fire-doors and all such doors must be hung so as to permit of opening from either side at all times. Near the centre of the highest part of the stage should be constructed one or more ventilators of incombustible material, extending at least ten feet above the stage roof and of an area equal to at least twelve per cent. of the area within the stage walls. Openings in these ventilators should be closed by valves so counterbalanced as to open automatically, and kept shut by cords, in which must be inserted a fusible link at a point near the bottom of the ventilator. This cord should be fastened on the stage floor level near the prompter's desk so that in case of necessity it can be easily reached and severed. All that portion of the stage not comprised in the working of seenery, traps and other mechanical apparatus and usually equal to the width of the proscenium opening, should be built of fire resisting material. The fly and tie galleries should be constructed of iron or steel, while the gridiron or rigging loft should have a lattice iron floor, and be readily accessible by iron stairways. All stage scenery, curtains and decorations and all woodwork on or about the stage should be saturated with some non-combustible material, and this should apply likewise to all finishing coats of paint given to woodwork. A strong feature should also be made of a careful and thorough examination before and after a performance of all fire apparatus by the firemen, whether municipal or private, attached to the theatre for professional duties. This in brief standardizes the main features regarded by experts as embracing the minimum demands consistent with the safety of the public and they have been given in some detail, since shorn of picturesque narrative, they are more likely to receive attention from the serious minded.

A better example of a calamitous fire at a theatre attended with appalling loss of life could not be selected than that of the Iroquois Theatre in Chicago. On the 30th of December, 1903, two thousand women and children crowded to a matinee performance of a musical extravaganza called "Mr. Bluebeard." The theatre had the reputation of being the largest, safest and newest in Chicago and had seating accommodation for 1,740 persons. Holidays were in full swing and being the last afternoon performance of the old year it drew hundreds of little ones with their happy parents not alone from the city itself, but from many towns in the vicinity. The house was consequently packed, many people willingly standing at the backs of the galleries in order to see the celebrated Eddie Foy, the chief laugh maker of the play. A particularly popular song was in progress, and children and grown-ups were absorbed in watching eight pretty girls and eight young men singing and dancing to the strains of the fine orchestra, when suddenly a large piece of burning muslin border fell upon the stage. Unknown to actors and audience the "spotlight" had fused, or so it was suspected, and stage hands with sticks had been fighting the fire for some moments in the wings before this ominous herald made its unwelcome appearance in full view of all. The singers gasped and wavered; the orchestra ceased with a crash. For the space of a heart-beat there was silence. Then a curious figure bearing a small child staggered from the wings to the footlights. It was Eddie Foy and the child was his son. Hurriedly he passed the boy to the conductor and the grotesque appearance of the comedian, clad only in his tights and minus half his grease paint, momentarily distracted the attention of the audience from the flames behind him. "For God's sake play and keep on playing," he implored the leader of the orchestra in hushed tones, and the musicians responded to his appeal with trembling hands and uncertain fingers.

Meanwhile, desperate efforts were being made to lower the fireproof curtain, which bellying in, owing to the draught from the auditorium, jammed and descended only a few feet. As the flames spurted from beneath its edge, a woman's shriek rang out and of the horror which ensued few of the survivors can bring themselves to speak. Fire and smoke driven from the stage swept up to the galleries, where a panic had already started. Mothers wrapping their arms about their children, were trampled under foot in the wild rush of despair. Then the stage loft collapsed. A column of flame rose from the ground to the ceiling and the theatre was plunged in darkness, while the battle of life continued in one crescendo of horror. All unconscious of the tragedy being enacted within a stone's throw, some painters in a building opposite one of the balcony exits suddenly saw a man standing on the escape. As they looked, the red glare of fire on the story below him showed them that his way to safety had been cut off and that his need was desperate. Running out a ladder from their window to where he stood they urged him to cross; but ladder and man slipped from the coping and plunged with a sickening thud into the street below. And now, more crazed creatures were making their way to this narrow platform and of them, twelve were drawn to safety on some planks. By this time, however, the fire was above, beside and beneath them, and women and children packed like sardines, helpless to move, were roasted slowly alive before the eyes of their would-be rescuers. When the firemen succeeded in entering the charnel house they were confronted by a wall of bodies ten feet in height and seven feet in width. It was impossible to believe that amongst these distorted forms could have remained any living person, but the Fire Marshal called to the surrounding silence, "Is there any one living here?" There was no reply, and the men made their way over this ghastly barricade in search of, perchance, one survivor.

Out of the two thousand merry-makers who had entered the theatre, buoyant with happy expectations, six hundred and two were carried to the morgue. It was found by the exploring firemen that those in the second balcony had suffered most in their futile efforts to descend the stairs. The sight was a horrible one. Wedged in a solid mass, which had practically lost all semblance of humanity, were what had once been men, women and children, twisted and entwined together in their death struggles. In the vise-like grip of usually feeble hands were found bits of cloth, fragments of jewellery and strands of hair evidently wrenched from their possessors in that hideous carnival of terror. One poor woman from being bent back over a seat had not only a broken spine, but had become practically dismembered through the pressure placed upon her, while in many instances faces had become so distorted as to be unrecognizable even to near relatives. Others who had bravely kept their seats and withstood the spur of panic fear had been overcome by smoke and gas and at least had received as their reward a peaceful death. A fire captain observing one of his men carrying the body of a girl called to him, "We've got no time for that sort of thing now, we must get on." To his surprise the man raised a tear-stained face and said brokenly, "Captain, I've got one of my own about this age, so if you don't mind I'll carry this little one out." The Captain silently handed him a blanket and, unrebuked, he bore his sad burden to the door.

Those there were, whose hearts were too hardened to be touched by the piteous spectacle, which had unnerved the strongest minded. These unspeakable creatures lurked and crouched in corners waiting for the opportunity to pull a ring from a powerless hand or to wrench a brooch from a motionless form. Over the scenes at the identification it is unnecessary to linger, suffice it to say that so widespread was the sympathy evoked by this terrible catastrophe, that for many days following Chicago was a city of mourning, all festivities being suspended.

The distressing incidents accompanying this outbreak have not been dwelt upon with the idea of satiating a morbid curiosity, but rather because they bring home forcibly to the general reader some notion of what a theatre panic really means. To how. many has the question ever occurred when seated at a theatre and enjoying a play, "What should I do if some one shouted 'Fire' now? Should I push and fight and struggle or should I remain calmly in my place?" It is this question of the personal equation which makes fire control in theatres a problem, at once perplexing and all-important. Obviously, the first step towards the safety of theatre audiences must come from a properly equipped and constructed building. Next, the records of almost every theatre disaster show that the critical moment in determining the fate of the audience is that immediately following the first indication of alarm. Hence the training of theatre attendants should be directed rather towards the prevention of panic than to the regulation of the movements of a panic-stricken audience. For one thing, the wide disparity in numbers between the available house force and the audience would clearly render abortive any attempt at such regulation. There are, however, well-defined principles, which if carefully observed will materially assist in directing the movements of an audience in case of fire. Firstly, to ensure the best results, all employees permanently connected with the theatre should be organized into exit drill companies, each individual member being assigned a special duty. While it is both necessary and important that the individual units of these companies should be instructed in the handling and use of fire equipment and grounded in the rudiments of fire extinction, the paramount consideration is the safety of the audience and every available means should be utilized of rendering assistance to the ushers whose business it is to obtain its prompt and orderly departure.

All fire signals should be transmitted by an electrically operated alarm system, the recording apparatus of which should be placed in the main business office, the box office and the stage manager's office. Upon receipt of an alarm by the stage manager, or when fire is discovered in the stage section before an alarm is struck, the curtain should be dropped immediately and the stage manager, one of the actors, or the fireman on duty should go before the curtain and announce the discontinuance of the performance. Upon the wording of that announcement and manner of its delivery will largely depend the behaviour of the audience and hence it is strongly recommended that a form of announcement should be prepared in advance and copies thereof placed in the hands of the various stage employees. It should be brief and to the point.

Something after the following is recommended:

"I am instructed by the management to announce that it will be necessary to discontinue the performance and immediately to dismiss the audience. Every one in leaving the house should implicitly follow the directions of the ushers stationed in each aisle."

Of course, the use of music, a lively march or something of a stirring character, is an excellent means of keeping an audience in hand and getting them away without unnecessary fuss or excitement. But this again presupposes an element of control over the orchestra, which it would be almost impossible to ensure, unless the same musicians were permanently employed and the management were satisfied that they could be relied upon to do their duty in case of emergency. Otherwise, obviously, they would be worse than useless and would probably only augment signs of unrest in the most undesirable quarters.

Another excellent method would appear to be to have a large plan of the theatre, with exits clearly marked, painted upon the fire curtain and exhibited for a few moments during each entre-act.

After the announcement of the cessation of the performance has been made from the stage, the ushers should move forward to their respective aisles and by word of mouth should quietly instruct their charges as to the speediest way to the street. For the assignment of exits the seat-plan on each floor should be divided into sections and to each section there should be assigned certain exits according to the relative discharging capacities, so that the time required for discharging the number apportioned to any one exit would average about the same for all. Each usher and doorman should be provided with a copy of the seating plan, on which should be indicated the exit assignments in detail. Ushers should, of course, be required to remain on duty by their respective exits during all performances. Fire alarm boxes should be placed in positions where they can be conveniently reached, but never in view of the audience. For the average theatre, there should be a box on each side of the parquet on the wall and in the rear of the last row of seats as well as one in the front hall. For balcony and galleries there should be also two boxes, one on each side behind the last row of seats. For the stage there should be one box on the rear wall, a box on each side near the proscenium wall and, when necessary, boxes in dressing rooms and the carpenter's shop. The boxes in the auditorium should above all else operate as noiselessly as possible, as a signal therefrom heard by an audience would be probably more productive of panic than even the sight of the actual fire. All theatres should be in direct communication with fire headquarters.

The system of assigning regular or pensioned firemen in uniform or of maintaining private firemen in uniform, where the regular force may be unduly depleted by such assignment to theatres, is to be commended. Their presence undoubtedly does much to inspire confidence and reassure an audience in moments of excitement, while naturally their superior knowledge and skill enables them to render valuable assistance when required. Finally there should be prominently displayed illuminated signs not only over the exits themselves, but in all conspicuous places, with arrows indicating the shortest and easiest route by which the street may be reached.

No doubt the writer will be told that he has suggested counsels of perfection and that if the caution practical experience demands in theatres, and which is embodied in this chapter, became law, theatrical managers would spend all their time in looking after minute details and audiences would resent being treated like children. Yet, as a matter of fact, in moments of crisis grown people are very often akin to children, which is evidenced by the fact that under such circumstances it is extraordinary how few otherwise level-headed persons will for one moment think of leaving a theatre by any other door than that through which they entered, quite irrespective of convenience of location. And hence it is that those in control must devise means to prevent them doing just those things which are worst for them, even at the risk of some unpopularity. And a fire in a theatre or a panic arising from an alarm of one, as has been shown, may lead to such ghastly results that it becomes the duty of all municipal governments to do all in their power to prevent such an occurrence. The writer, without wishing to appear extreme, is of opinion that some limitation should be placed upon the seating capacity of theatres as distinct from stadia and places of that nature. An audience of 1,800 is sufficient to tax all the resources of those responsible in emergencies and is about the maximum number which can conveniently and quickly pass out of any theatre without causing untold confusion in the street, which will in its turn hamper the fire forces. Equally, however, a theatre run along the lines indicated will not only promise the maximum of safety, but without exaggeration will afford its patrons a greater amount of security than as a rule they will find in their own homes.

CHAPTER XIV

THE decisive feature governing fire-fighting in all countries and under all conditions may in every case be summed up in the two words "water supply." Personnel may be of the finest, and apparatus of the most complete, but both are helpless if the wherewithal to quench the fires is lacking. Many disastrous conflagrations have owed their magnitude to this circumstance, and it is a curious commentary upon municipal intelligence that many large cities the world over, surrounded as they are with an abundance of water, absolutely lack means for concentrating it at the scene of a serious outbreak. Small mains intended for supply under normal circumstances become practically useless when great fires are in question. Further, fire departments are often criticized by the inexperienced in newspapers and elsewhere, for their inability to check a blaze when the fault really lies with indifferent "city fathers," who, in their omnipotence disregard the advice of those, who after all are paid to know, and absolutely fail to benefit by past bitter experience. During the year 1903, several large fires occurred in New York, and several disastrous ones of great magnitude throughout the United States. The city had been growing steadily, it was recognized that the water mains were too small to meet an emergency, and the authorities thereupon decided to investigate this most important subject. After careful discussion and consultation with eminent engineers, it was resolved to install the most up-to-date system of water supply known to science, popularly known as the "high pressure" service. In its essentials there is really nothing very remarkable about the idea, and in fact its designation as above is something of a misnomer. As a matter of fact the actual pressure off a "high pressure" main can be equalled by the modern steam fire pump. But, whereas the latter is dependent upon the human element in the shape of the fireman who is responsible for the stoking of the engine, for the quality of the coal and the organization of the fuel supply, and finally for climatic conditions which, in extreme circumstances must affect in some degree an unprotected boiler exposed to the fury of the elements, this alternative system is practically independent of all these considerations. Undoubtedly the ideal situation for such an installation is in a town which draws its water from surrounding mountains, such as in many Swiss cities. It stands to reason that if a reservoir or lake lies some thousands of feet above the point to be supplied, the laws of gravity will insure a steady and continuous stream of water at any position in the area connected with it by mains at a pressure according to the altitude of the source of supply. The fire departments of Switzerland have shown themselves keenly alive to these natural advantages, and the mechanical fire pump is practically unknown in their fire departments. But, however, the world has not been formed for the convenience of its occupants, and hence it is that science has been compelled to step in and, by artificial means, to find the solution of the problem. A brief description of the "high pressure" system written from the standpoint of the fire-fighter will explain the scheme of operations, and what applies to one city applies, to all intents and purposes, to others.

The service in the Borough of Manhattan—the island of Manhattan in the city of New York—protects approximately 2,600 acres; that in the Borough of Brooklyn, about 1,400 acres, and that at Coney Island about 146 acres. There are two pumping stations in Manhattan, with 2,066 hydrants, and some 300,000 feet of mains, chiefly located in the business section of the city. Brooklyn

CAPACITY TEST   HIGH PRESSURE SYSTEM   NEW YORK

has 1,112 hydrants, including 24 for fire-boat connections, while Coney Island possesses 345, including three Monitor nozzles. In deciding upon the location of pumping stations, prudence naturally directed that they should be placed so as to be practically outside the reach of any possible conflagration, and yet in a position to avail themselves of an unlimited supply of water drawn from either fresh or salt water sources. Thus the Manhattan stations were located at the northwestern and southern ends of the protected area, the main features of their construction—one-story and basement, fire-proof buildings—being almost identical in both cases. These structures are of sufficient size to carry eight pumping units each, though the present equipment consists of but six. The contract calls for a delivery from each pump of 3,000 gallons of sea water per minute against a discharge pressure of 300 pounds per square inch, and a suction lift not exceeding twenty feet. At the acceptance tests, the fire pumps in each station totalled a delivery of about 18,000 gallons per minute at the aforesaid pressure, some of the individual pumps discharging as much as 3,800 gallons. This total can, of course, be increased proportionately without change in the buildings or mains, by the addition of the two pumping units for which space has been provided. Fresh water for each station is supplied through two twenty-four inch mains connected with a third of thirty-six inches diameter.

The salt water supply is drawn from the North and East rivers through two thirty inch pipes. These lead into suction chambers directly in front of each station, and are so constructed that they are at all times below mean low water. This ensures a steady flow and prevents the possibility of interruption caused by air being admitted to the suction lines. Protection is afforded to the river ends of these mains by heavy bulkhead screens, and to the suction chambers by weighty bronze shields which are readily accessible for cleaning. The pumping units consist of centrifugal pumps driven by electric motors, both supported on a common bed. Special care as to strength and ability to resist corrosion was expended upon the pumps which run at a speed of 740 revolutions a minute. They are of the five-stage type, each stage being designed to give a pressure of sixty pounds to the square inch, or the combined pressure of the five stages, three hundred pounds to the square inch, which is the maximum working pressure of each unit. It may sound a scientific anomaly, but the fact remains that increase of pressure does not correspond to increase in volume. To frame a crude analogy. An "ocean greyhound" can steam eighteen knots at an economical coal consumption. Increase her speed two knots and the consumption of fuel at once increases out of all proportion to the additional speed. Subsequently, each additional knot, or even half-knot, will demand an enormous increase in coal consumption, till eventually a certain maximum of speed will have been reached beyond which it is impossible for the engines to develop sufficient driving energy, no matter how much coal be expended. In other words, thereafter, surplus energy becomes waste.

Now, somewhat similarly, each installation of pumps can deliver about 18,000 gallons per minute, at a pressure of three hundred pounds, but a much greater volume of water can be secured by running them at a lower pressure. Thus the average pressure required for fire duty is from 125 to 200 pounds at the hydrant, and each station working under these conditions will deliver with its present complement of pumping units, 30,000 gallons of water per minute. Were this volume of water concentrated within a radius of five hundred feet, no imaginable conflagration could survive its attack. The pumping units are set in operation by throwing a switch on the main switch board, which is directly connected with the motors. By this means the machines are brought into instant use, and in less than one minute the maximum pressure can be developed. Current for these motors is furnished locally at a pressure of 6,600 volts, and each station has four separate electrical feeders, two from the waterside station and two from the nearest sub-station. Of these four feeders, two will operate the six pumps, but provision is made for connection with the Brooklyn stations of the electric company in that borough, this in cases of emergency. Two twenty-four-inch mains lead out of each station and traverse practically the entire protected area; these are intersected by lateral branch pipes of twelve and sixteen inch diameter, which, in turn, are cross-connected by twenty-inch mains, at frequent intervals, the water thus traveling only a short distance through a main smaller than twenty inches before reaching the hydrants. These latter are connected to the mains by eight-inch branch pipes, gates being provided at intervals of about 250 feet to enable the carrying out of necessary repairs without affecting any hydrants except those directly adjacent to the gate in question. The pipe system is so planned, that without excessive drop in pressure due to friction loss in the mains, it is possible to concentrate 20,000 gallons of water a minute upon the average block of buildings, or the full capacity of both stations upon an area of approximately one-quarter of a square mile. Since the work of the fire department commences at the hydrant, no part of a "high pressure" system is more important in determining its efficiency than the type of hydrant employed, which must be ever ready for instant service.

Hence, the following points may be tabulated as governing the selection of this most essential feature. Firstly,—suitability of design; this includes workmanship and material in order to obtain the maximum of reliable service. Secondly,—facility of operation. Thirdly,—freedom from frictional resistance, to ensure maximum delivery; and fourthly, and of supreme importance,—perfect drainage to obviate all possibility of freezing. Without going into a lengthy dissertation upon the specifications prepared and the tests carried out, it is sufficient to say that the hydrants eventually installed are furnished with four three-inch outlets, which provide a capacity of four two-inch streams with seventy-five pounds nozzle pressure, and roughly four thousand gallons of water per minute. It was deemed advisable that the main valve in these hydrants, which is six inches in diameter, should open downwards against the pressure to be encountered. In order, therefore, that the operation of opening this valve against the heavy pressures liable to be met with in the service mains should be an easy and rapid one, it was so designed that the first three turns of the hydrant wrench should open a pilot valve, thereby admitting water to the barrel of the hydrant, thus equalizing the pressures on both sides of the main valve, after which it can be opened without difficulty or resistance. A few seconds is sufficient to accomplish this adjustment of pressures after the application of the wrench.

The operation above described sets in action a drip valve, which closes as the main valve is opened and opens as the latter closes. There is a connection between the drip valve and the sewer, thus ensuring the drainage of the hydrant barrel after use. Valves controlling the hose outlets are provided with a device which balances the pressures, permitting the former to be opened easily with a five-inch wrench under a pressure of 250 pounds. Needless to say, before this system was finally handed over for practical service in the New York fire department, extensive tests were made, two of which are particularly worthy of notice. The first took place along the North river front. Twenty-one three-inch lines were stretched from seven hydrants, and twelve two-inch and nine one-and-a-half-inch nozzles were used. Within two minutes after the order to start water had been given, a nozzle pressure of eighty pounds was registered, and so great was the volume of water delivered that the streets speedily became a lake and overflowed towards the docks. The second test was even more exhaustive. Twelve three-inch lines with one-and-a half-inch nozzles, six siamese lines with two-inch nozzles, a water tower with a two-inch nozzle, and a deck pipe with a one-and-three-quarter-inch nozzle, were all brought instantaneously into action. One minute after giving the order to start, a nozzle pressure of 150 pounds was obtained, and in two minutes 195 pounds was registered on the one-and-a-half-inch nozzles, and 170 pounds on the two-inch. With the nozzles elevated to an arc of eighty degrees this pressure carried a solid stream of water one hundred feet above the roof of a fourteen-story building. The gauge on the water tower at this time registered 270 pounds. To the layman, even these simple figures may seem perplexing, so it may be well to translate hard facts into picturesque simile. A stream of water propelled by a pressure of two hundred pounds to the square inch striking an ordinary office partition at right angles at a distance of about one hundred feet, would smash it up as though it were so much matchwood and play with its contents like a whirlwind. Similarly, the same stream, elevated to eighty degrees, could easily clean the cornice of the tallest apartment house in Manhattan, or could wash the dome of St. Paul's Cathedral in London.

In 1908, the system was formally turned over to the Department, but at the outset the greatest caution was observed over its operation. Thus, companies responding to an alarm in the high pressure district, as a precautionary measure, always coupled up to the low pressure hydrants with engines, thus making security doubly sure. As the absolute reliability of the new installation became increasingly apparent, these precautions were gradually withdrawn, till today no engines respond to any fires in the high pressure district, unless specially requisitioned.

Many outbreaks now-a-days are fought and conquered with a second alarm assignment, which, before the advent of the high pressure system, would have required a fourth alarm, a fact which, trivial perhaps to the lay mind, will be appreciated by Chiefs of Fire Departments in its full significance. No doubt, the latter have often realized the danger of having to draw almost every piece of apparatus at their command to a big fire, leaving nothing in reserve with which to tackle another outbreak should one then occur. Simultaneous calls of this nature tax the wit of man to meet, and the days of miracles are past. The first fire can be fought and quelled, but the second, coming when the force is already engaged, forms a very serious menace. In the high pressure area of New York, four fires of some magnitude have been attacked and suppressed at the same time, not to mention several smaller blazes. That fact speaks for itself. Prior to the installation, also, there was always the fear of a sweeping conflagration after the style of Baltimore; but that has passed away forever. New York today is the best fire protected city on the American continent, but even the knowledge that this is so has not prevented further efforts to meet the demands of the situation. Another pumping station is even now in course of construction and within two years it is confidently expected that the area protected by duplicated high pressure mains will amount to over ten thousand acres, this in the most congested and most valuable portion of the city. But to return to the detailed description of the apparatus employed in rendering the installation additionally effective.

The use of centralized energy for delivering water at a high pressure throughout an entire system naturally entails upon all the mains and hydrants involved the maximum pressure required at any one point. It is, therefore, quite likely to happen, that whereas a pressure of 125 pounds will be ample for one fire, another outbreak may occur in the protected area necessitating a pressure of 250 pounds. This would mean that the pressure would have to be raised suddenly to meet the fresh call, and, coming unexpectedly and unheralded, it might seriously endanger the men operating at the smaller outbreak. Or, further to illustrate this point, it is sometimes desirable to take water from a hydrant for a hand line at seventy-five pounds, and from another outlet to supply a water tower at 225 pounds. To permit of this arrangement, a regulating valve, weighing only twenty-five pounds has been invented, which is attached to the hydrant outlet. A pressure gauge is inserted on the hose side, the regulating valve is opened until the gauge needle points to the pressure required, and no matter to what extent the pressure on the receiving side is increased, on the discharging side it remains at the figure selected. If a shut-off nozzle is used on the line, automatically the pressure remains at the figure indicated on the gauge. Even the layman will appreciate the enormous importance of a valve of this nature, without which the system as the whole could not be properly controlled or operated to the best advantage.

In order to guard against the somewhat remote possibility of a breakdown of all pumping units in both stations, provision is made for connecting the fire-boats of the department to the high pressure hydrants. By this means these boats can deliver 60,000 gallons of water per minute at a pressure of 250 pounds. In addition, by this method, fire-boats may be made to constitute a valuable auxiliary in the case of any great emergency. But, as science is never idle and one discovery leads to another, so it is with the business of fire control, and day by day fresh ideas are brought into being, tested and utilized for the common, weal. Thus, in the later portions of the city protected by the high pressure system, it has been found advantageous to duplicate the mains and hydrants, so that if one set of mains becomes blocked, the other may be brought into service without delay. The system is operated as a unit, but in case of a break in the mains, it can be divided by the closing of the motor valves at selected points. These valves are controlled by switches on the pumping station's switch board, a row of red and a row of green lights indicating the two lines of mains. In the event of a break, pumps are shut down and the motor valves are immediately closed. When the pumps are started again, the pressure building up on the line that is intact will be shown by the colored lights mentioned. The serviceable main will be kept in use at the fire, but the officer in charge must wait about one minute in order to see whether the pressure is from the red or the green line. As soon as the pressure shows on either, those lines connected to that disabled will be shifted over to the hydrants on the main intact and the fight can be continued with ample water supply.

During the continuance of a fire, the Chief operating can control the pressure at the pumping station, increasing or decreasing the same or stopping it altogether by means of a special telephone service run into boxes conveniently situated and specially designed for the purpose, which are plentifully installed throughout the protected area. This system has been found to answer well, though it is open to the possibility that orders may be misunderstood from one cause or another. Probably some improvement thereon may be framed in time as the science of fire-fighting becomes increasingly popular and it is better realised what an important part is played by fire control in the daily life of the people. For instance, one of the most gratifying effects of the introduction of the high pressure system in New York was the immediate drop in insurance rates. In December, 1908, about six months after this system had been put into regular use, the New York Fire Insurance Exchange made a general reduction of rates throughout the high pressure zone in Manhattan, a reduction amounting to the respectable sum of $500,000 annually. In Brooklyn, likewise, a reduction of $250,000 followed, so that the improvement has already saved the taxpayer in the community about $750,000 a year. Hence, this method is not one of fire control alone, but becomes positively a good investment.

Of course, it may be urged that in the United States, the profession of fire-fighting has developed into such a highly specialized service and has demanded so many drastic changes in equipment and maintenance, because the risks incurred are proportionately greater than are met with elsewhere. There is a measure of truth in this, but it is on a par with the individual, who, living in a small island, considers motor cars an unnecessary means of transportation and railways a wasteful expense, because he has always found a horse and trap convenient and sufficient for his needs. That is, of course, a somewhat exaggerated simile, but it is an incontrovertible fact that there are many otherwise normal citizens in all the countries of

SWITCH-BOARD AT HIGH PRESSURE PUMPING STATION.

the world who view innovation with suspicion and shield themselves behind the comfortable assurance that "our fathers did very well without all these new fangled notions and what was good enough for them is good enough for us." This especially as regards municipal outlay on fire protection, though in this respect the writer must lay it on record that New York has always risen to the occasion in the most openhanded manner. For some obscure reason the average man objects to putting his hand in his pocket for anything in connection with the fire risks of the community just as much as though he were being held up by a highway robber, and willy nilly had to surrender his purse.

There is no advantage in labouring an obvious point and the writer has in his mind not Europe alone, but some American cities where the Fire Department is apparently regarded as a costly and unnecessary adjunct to the other municipal offices. Yet in all fairness to those whose responsibility it is to protect the lives and property of those given into their charge, there should be no hesitation in providing them with the most modern and up-to-date apparatus for the discharge of their duties. The mathematical aspect of the whole question as regards the "high pressure" system as a whole, is not a difficult one for Fire Committees to understand. Since one "high pressure" hydrant is the equivalent of six engines of the first size, how many hydrants would be equal to the entire pumping outfit of the brigade, using an English term, and would not the initial expense of installation be more than met by decrease in personnel, decrease in insurance rates and increase in safety of citizens committed to their charge? The answer to this conundrum should not unduly tax the mental equipment even of one of those corporations addicted to the traditions of the Medes and Persians.

CHAPTER XV

SINCE it is no exaggeration to define panic as one of the most effective allies of fire, it is obvious that in dealing with buildings occupied by large numbers of either young or infirm people, or with places where passing crowds are apt to congregate, such as department stores, peculiar precautions are necessary. The genesis of many a conflagration, attended afterwards with terrible loss of life, is often trivial. Taken in time and dealt with coolly it would never have developed into a serious outbreak, and equally the magnitude of the blaze, as regards actual fire damage, can never be accurately gauged by the death roll. Experience has shown that in such disasters as many die from suffocation consequent upon crushing, or from injuries received in seeking safety through some desperate and ill-judged action, such as jumping into the streets, as perish in the flames. Of course, this is natural. A curl of smoke, a few sparks and a cry of "fire," and unless beforehand prepared for this kind of emergency, the primal impulse of any one is to reach safety, or what appears safety, as quickly as possible. No thought is given as to the best mode of exit, misguided instinct suggests the way by which one has entered, and instantly corridors, stairways and passages become jammed with a frightened, hustling crowd, beyond control, and following each other like sheep to the shambles. It is this incontrovertible fact which has caused the architect to labour towards the design of "panic-proof" structures, and has led those interested in fire control to devise means which shall render such occurrences rare to

the point of non-existence.   And it must be remembered
that unfortunately, responsible authorities are called upon to
frame regulations, oftentimes not for application in modern
fire-resisting buildings, but for structures composed largely
of lath and stucco, and which, given the opportunity, would
burn like tinder boxes.   Hence, it seems scarcely necessary
to emphasize the point that, with the latter, the danger is
more acute, and above all else there is need for speed and
sang-froid.   Now the only method by which this state of
affairs can be assured is by a process of accustoming the
human unit to the conditions likely to arise in a fire emer-
geney, and this can best be done by means of drill.

Whether it be employees in a factory, children in a school
or the staff of a hospital or a department store, exit drills
should always be enforced, and where circumstances allow,
provision should be made for some sort of "house" fire-
fighting force.   It must be understood, of course, that the
latter in no way takes the place of the regular fire department, which should be communicated with at once in all
cases, but rather is intended to act as an auxiliary pending
their arrival.   It will be convenient to deal *seriatim* with the
four types mentioned.   In factory buildings particular danger attaches to the stairways connecting stories.   They can
only accommodate a limited number of people and form
dangerous exits for crowds.   Congestion at their corners
means death, and since employees may be expected to vary
in nationality, misunderstanding of orders becomes more
probable, and the problem of preventing a panic assumes
a thorny aspect.   In the first place, exit drills should be
held as often as possible and should include every one in
the building.   When two or three firms occupy the same
premises there should be coöperation, and the alarms announcing these drills should be given from different floors
in order that practice may be afforded in changing the order
of precedence for possession of stairways or fire escapes.
The line of march may be so arranged as to take advantage
of the additional time required in the descent of those from
the upper floors, by dismissing such of the lower floors as

would not delay their egress. An exception to this rule should be made where buildings are divided by fire walls with protected openings, which permit of the transfer of occupants in the fire section to the corresponding fire-free section on the same floor in the same building, or where provision is made for a safe retreat by means of gangways leading to adjoining buildings. Incidentally, an excellent scheme, where feasible, is to dismiss the employees nightly by a fire signal. In assigning stations, the first consideration is the selection of aisle guards, whose business it is to effect line formation, prevent pushing and overcrowding and to see that the time honoured precept "women and children first" is observed. All subsequent movements should be regulated by a gong or whistle. Thus, the first alarm indicating the floor of the outbreak should consist of a number of taps indicating the floor. As soon as the first stroke sounds, work should cease, and, if possible, all power be shut off the machines. Then all stock, chairs or benches blocking the aisles should be removed by the employees nearest, and placed either above or below the work tables. The next movement is to march to the exit passage in single or double file. If in the latter, couples should link arms for mutual support, the women using the free hand to raise their skirts to prevent themselves and those behind them from tripping, and each file should move forward, observing a uniform distance between the couples. The signal to start should be given by the "room captain," and under no circumstances should any employee be permitted to attempt to secure clothing from locker or cloak room. Upon reaching the street the line should be led away to a safe distance from the building, and for this duty one of the supervisors should be selected and drilled as a guide. Elevator attendants should take their cars, upon the first sound of a building alarm, to the floor indicated, and hold themselves subject to the orders of the "floor chief." In high buildings of the fire-resistive type, the operator should run his elevator into the fire zone, receive passengers, and if conditions favour, discharge them a few floors below.

The usual difficulty, however, is that floors and stairways are so crowded that he has no option but to run to the ground floor. The assignment of exits necessarily depends upon their number, capacity and location. But it is important that all means of egress should be based on approximate estimates of their relative discharging capacities, which can be readily arrived at by actual tests.

When possible, provision should be made for the unhampered entrance of firemen, and in the planning of such fire drills combinations of exits should be studied. Employers having the welfare of their work people at heart, can always obtain advice, if in any doubt on this subject, from officials of the fire department. All that is intended here is to suggest certain simples rules of conduct which will tend to prevent confusion and make for safety. The location of stairways, fire escapes and other exits should be indicated by illuminated signs, and for the information of employees, leaflets should be printed in several languages, giving the details of the fire drill. It goes without saying that all modern buildings of this type should be equipped with an electrically operated alarm system, the mechanical gong of which could be better heard above the noise of any machinery than one struck by the "room captain," and would possess the additional advantage of automatically operating on all floors from any position. The box stations governing these alarms should be accessible only to responsible persons. It may be urged that such precautions presuppose certain members of the staff being possessed of intelligence and a considerable amount of organizing ability, but as a rule either foremen or forewomen in sectional charge of fifty employees will be found to fill admirably such executive positions as "room captains." They in their turn are naturally subordinate to the manager, who should accept supreme control and the responsibility attaching thereto. Aisle guards may be compared to lieutenants, should be strong and alert, and owing to the fact that they may be required to use some physical force, should, when possible, be men. They should be especially watchful of persons stumbling, fainting or becoming hysterical. Where stair exits have sharp bends, they should be stationed there to prevent congestion, and above all, they should be made to realize their obligations and to feel that their duties are no mere sinecure. Finally, there should be at least one male and one female searcher on each floor to visit the toilet rooms and other such places, where perhaps the fire signal cannot be heard. In buildings of an antiquated type these precautions make no pretense of securing absolute safety to the individual, but if the drills be arranged with the advice of a skilled fire official it is probable that panic will not unreasonably seize all inmates, and that the fire department will at least be given an opportunity of effecting such rescues as human ingenuity and providence will allow. At any rate they will not be met with the appalling conditions which alas, have been only too common hitherto, when persons have found death needlessly, while safety awaited them with the advent of the professional fire-fighters. The fire risk is real enough without the additional factors of fright and bad management, and it is to guard against these that the above has been suggested.

One of the most terrible conflagrations of this nature in recent years was that which occurred on Saturday, March 25th, 1911, in the Asch Building, a ten-story structure situated at the corner of Washington Place and Greene Street, New York City. The following account, vivid in its simple realism, is taken from the report of the New York Board of Fire Underwriters.

"Occupancy, 8th, 9th and 10th floors. Work rooms, show-room and factory, stock-room, pressing and shipping department of the Triangle Waist Company. . . . On the 9th floor there were two wooden partitions, one forming the cloak room, the other being at the north side enclosing the entrance to the freight elevators and stairshaft. .

On the 10th floor there were partitions of wood and glass forming offices and show-rooms.   . On the rear of the building in the court was an iron fire-escape, the steps being 17½ inches wide. The fire-escape did not extend to the bottom of the court and the latter had no exit to the street. On the eighth floor there were five unbroken rows of four-foot tables, each containing a double row of sewing machines and shirt waists in process of manufacture. These tables extended from the Washington Place front 'south wall' to within eighteen feet of the north side of the building. This latter space was partially filled with stock, principally on tables. An aisle space was also left running east and west along the north side. The space along the east wall contained the cutting tables. Approximately 275 operators were on this floor. On the ninth floor there were eight unbroken rows of four-foot tables with 300 operators. There were no aisles running east and west at the south side of these floors, the sewing machine tables extending close up to the wall. The space between the tables was approximately four feet wide and contained two rows of chairs back to back. It also contained baskets and other receptacles for the goods in process of manufacture. The only convenient way for the operators next to the south wall to reach the stairs and elevators at the southwest corner, was to walk the entire length of the crowded space between the tables to the north side, and then use the aisles which extended along the north and west sides of the building.

"The fire started at 4:42 P. M. on the eighth floor in the vicinity of the northeast corner of the building, almost simultaneously with the signal to stop work for the day. It is generally believed to have originated from a match or cigarette igniting scrap material on the floor in the vicinity of the cutting tables. It spread rapidly, however, due to the large quantity of inflammable material consisting chiefly of thin cotton, lace and other trimmings. In a very short time the fire had spread over the entire floor and communicated, principally out and in the windows, to the floors above. In addition to the windows, the fire may have communicated from floor to floor by way of the stairs and elevator shafts, as the doors were undoubtedly open, in part at least. The plant was working overtime when the fire occurred. According to the information obtainable the operators crowded in among the machines, chairs and goods on the eighth and ninth floors, were badly panic stricken immediately after the start of the fire, and in consequence made slow progress towards the exits. Considerable delay is said to have been experienced in opening the doors leading to the stairs at the southwest corner of the building as they opened inwards and the women became jammed against them. Practically the entire loss of life was confined to those employed on the ninth floor. More than half of the number said to have been on this floor escaped. It seems apparent, however, that by the time this number had got out, the elevators had stopped running and the flames around the two inside stairways and outside fire-escape, both on this floor and those adjoining, would not permit any further egress in these directions. The result was that all who remained on the floor until this condition prevailed were overcome by the smoke and fire or jumped from the windows. It is said that a few—probably twenty—from the upper floors descended by way of the outside fire-escape. These reëntered one of the lower stories and passed down the stairways. Approximately twenty-five bodies were found closely jammed in the cloak room, next to the stair shaft at the west end of the building. About fifty were found near the northeast corner behind the partition and clothes locker located thirty inches from the north end of the two tables nearest the east wall. Twenty bodies were found near the machines where they worked, apparently having been overcome before they could extricate themselves. About ten are said to have been taken from the bottom of the court on the north. The balance of those killed—approximately forty—jumped from the windows to the street."

There seems to be no doubt that had fire-drill been organized amongst these women so great a panic would have been avoided. But in the opinion of the writer automatic sprinklers would in all probability have averted the disaster

WATER TOWER AT WORK.

as 'their operation would have turned in an immediate alarm and the delay in sending in an alarm contributed greatly to the appalling loss of life. It must, however, be remarked that the fire-escape in the rear of this building was quite' inadequate for the needs of the situation, as in order to gain the street, it will be noticed, that those using it were obliged to reënter one of the lower stories and pass thence down the main stairway to the front door. In addition, doors should never be constructed to open inwards.

Turning now to the problem of schools; in its essentials this is in many ways akin to that of factories, with the outstanding difference, that in dealing with children even greater care must be exercised by the supervisors or "room captains." These should be chosen from amongst the teachers, and their duties with regard to their charges should be along precisely similar lines to those already laid down, though it should be borne in mind that personal influence here plays a greater part. Where pianos or other instruments are available an excellent plan is the use of march-time music to assist in steadying the lines of scholars after a fire alarm. Incidentally, school should always be dismissed once a day in accordance with the practices of fire drill. In the matter of exits, preference should be given to the classes of smaller children, and it is particularly urged that exits for infants should be smoke-proof and of sufficient width to accommodate double lines of two children each. Further, as far as the construction of the building will allow, the convergence of two columns in narrow halls or stairways should be particularly avoided. This is only too liable to cause confusion, which in the event of the building being a quick-burner, may result in terrible loss of life. In schools of advanced grades, where there are boys of a certain age, it is a good system to organize a small fire-fighting force to use the chemical extinguishers common to all public institutions, and from the nature of the duty, youngsters are likely to become enthusiastic over, and expert in, their management.

Since example is proverbially better than precept, the following accounts of two school fires, widely differing in ultimate results but having many points in common, may be not without interest to the general reader.

In Collinwood, a suburb of Cleveland, Ohio, there stood on the 4th of March, 1908, a large school accommodating over eight hundred children. The day was a warm one and there was but a small fire in the furnace, which was situated under the front stairs. Before the noon recess the janitor in charge noticed a thin stream of smoke coming from the basement and at once gave the alarm. On the ground floor the children were marched out quietly, calm in the belief that the signal was for drill, but before the anxious teachers on the next two floors could marshal their charges the fire had gained such ground that all escape by the front door was impossible. As the children neared this exit they were driven back by the smoke which confronted them and fought to reascend the stairs, only to be pushed down into the flames by the excited and frightened little mob still descending. In the drills used, both teachers and children had been accustomed to employ only the front door as a means of egress, and the fact that this means was debarred them seems to have had a paralyzing effect upon all intelligence and action. By the time the second stairway leading to a door in the rear was thought of, the children were entirely out of hand, and when this door was found to be locked the situation became uncontrollable. Parents brought to the scene by the sight of the smoke and the shrieks of the children in distress stood helpless, as did the firemen. In this suburb the only force available was that of volunteers, whose apparatus was inadequate, having no ladders long enough to reach the third floor and who were unable to obtain sufficient water pressure to extinguish the fire in the second story. One desperate mother, aided by an unknown man, tried vainly to open the rear door, behind which muffled sobs and groans told of the extremity of the little ones within. But her efforts were fruitless, and with her bare hands she succeeded in breaking some panes of glass in adjacent windows and managed by this means to drag to safety a few semi-conscious tots. None of the children fighting and struggling for life behind this pitiless barrier was more than fourteen years of age, and many were only six and seven. At the front door the weight of human bodies became so great that it collapsed and showed to the agonized spectators, many of them parents, a heap of little forms caressed by the flames and half hidden by the smoke. Amongst this pile was one small girl of ten, whose father arrived in time to make a futile attempt to pull her from the death awaiting her. Still alive, but crushed and horribly burnt, she was able to hold out her feeble arms to him, and he, heedless of the peril of his own position and intent only on the saving of his daughter, worked frantically until his own injuries prevented further effort. Another child was recognized by her mother. Their hands met, when a piece of broken glass fell on the mother's wrist practically severing it from the arm. As the grip of the two hands relaxed the daughter fell back into the blazing pyre to be seen no more. In thirty minutes from the time of the first alarm nothing remained of the building but four blackened and uncovered walls and a smouldering heap of wreckage, some of which had once been human beings.

It was only then that the firemen were able to enter the ruins, and there was virtually nothing for them to do. Of the 810 children who had taken in their books that morning, about 170 had perished, and with them had died two teachers in the vain attempt to lead their charges to safety. The rear door, also, was broken down by the number of little ones who had been packed so closely against it that their combined weight caused the lock to give, when too late. Practically all the bodies were unrecognizable, and frenzied relatives were unconvinced of the losses in their homes until the roll had been called. The origin of the fire still remains unknown; it may have been due to defective flues or to carelessness, but, be that as it may, the results of this catastrophe carried mourning into hundreds of homes and once again emphasized the pressing need of every known structural precaution in such buildings plus better considered planning of drills.

It is a relief to turn from the recitation of such horrors to the narration of a brighter and happier story. In Raleigh, a small town in North Carolina, on Friday morning, February 14th of the present year, a fire broke out in an old wooden school building, which, from its construction was a veritable fire-trap. In spite of the fact that the halls and rooms were filled with smoke before the 350 children could be got into line, their order was unbroken and their courage unshaken as they marched through the suffocating atmosphere to the doors and down the wet fire escapes. The principal of the school was notified of the danger by one of her subordinates; quickly closing the doors in the upper hall she gave the signal for the drill which her pupils had often practised. At the tap of the gong every child fell into line, those downstairs going out of the front and main entrances, while those on the upper stories descended to the streets by the two fire escapes, which were wet and sticky from the snow that had fallen during the morning. All was as orderly as a stage rehearsal. Even the smallest tots followed the elder ones without the slightest confusion. There was no attempt to get hats or wraps or books. The whole operation occupied only three-quarters of a minute, which was better time than had ever before been made in practice. Parents who had rushed to the scene, dreading the terrible sights which might meet their eyes, saw an orderly procession of youngsters march out of the building, filled though it was with smoke and flames. Owing to the snow and the slippery condition of the streets, the fire department had been appreciably delayed in responding to the alarm, and had it not been that the Fire Chief had insisted on and enforced the precautions of daily drill amongst the pupils, the loss of life might have been appalling. In the opinion of the writer, all concerned deserve the maximum of praise; the head of the Fire Department, Sherwood Brockwell, a graduate of the New York Fire College, for his insistence, the Superintendent for the intelligent way in which the children had evidently been trained, and the latter for their coolness and evident trust in their teachers. It is no exaggeration to say that the fire peril could be practically eliminated in schools, were the example of Raleigh followed. At the same time it is absolutely incomprehensible how sane persons, ignorant though they might be of the elementary principles of fire control, could allow so glaringly foolish an arrangement to be made as that which permitted the placing of a furnace immediately under the front stairs of a school building, apart from these having been constructed of wood. Under any circumstances, heating apparatus of that nature should be located in a separate structure adjacent to, but isolated from, the school itself.

Should fire occur during school hours the officer in command of the fire force can ascertain quickly from either the Principal or teachers the location of the fire, which will govern his subsequent actions. Should it be in the upper floors the entrance must not be attempted by doors or stairways by which the children are leaving, though use may, of course, be made of any unused stairway. By a ladder raised to an upper window a line of hose should be quickly brought to bear on the blaze, care being exercised to drive it back from the exits. Other lines as necessary will be placed similarly and the fire thus held in check till all the scholars are out of the danger zone. Then, should the fire assume dangerous proportions, it may be fought as in other buildings, i. e., by stairways both front and rear, and if necessary from both sides.

Too much emphasis cannot be laid upon the necessity of preventing excited parents and others rushing to the entrances by which the children are leaving the building, breaking the line, causing confusion and retarding the exit of those still within. However good their intent, their interference must work mischief. It is imperative that the children be kept marching until all are safely out of danger. Officers in command should see that an adequate force of police and firemen are told off for this important duty.

The safety of the shopping public in the enormous department stores, which have latterly sprung up in all American cities, and for that matter, in Europe as well, is in itself one of the most difficult problems which those interested can possibly face. Here there is no question of drilling regular habitues, for the population is a floating one, that is to say, the attendants and employees may be trained till they are expert in the duties assigned to them, but the dealing with, and dispersal of, great hordes of strangers is one that requires almost superhuman management and foresight. In fact, in the opinion of many it is impracticable; drills are rendered difficult by the constant presence of strangers, a test alarm may produce a panic, when even those gifted with the maximum of human magnetism would find the control of strange crowds beyond their powers. But, at least, precautions can be and are taken.

Abroad, as well as in the United States, efforts have been made by private fire departments, which should always be captained by retired officers of regular fire brigades, and by the organization of all employees into a homogeneous unit of action, in the event of crisis to grapple with the events likely to occur so far as circumstances will allow. No human agency can do more. The study of scientific fire control is of recent growth, and many of the great emporia which dot the cities of the world are the result of evolution, but as a rule when additions are made to such structures, they are subjected to the most searching of fire tests and the writer can aver from personal experience in New York that neither time nor money has been spared to render the same as secure as is feasibly possible. In addition, it must be understood that under no circumstances can effective drills be carried out, i. e., as though under emergency conditions, unless the temper of the public changes in an amazing degree. However, it is always practicable to construct some sort of edifice even upon the most insecure of foundations, and certain primary precautions though in no way adequate properly to control the situation, may go a long way towards the prevention of a disastrous panic.

In brief, all that has been written may be taken as supplying in embryonic form the basis of department store exit drill; that is to say, there should be capable "floor masters," capable "guards," and by private instruction the actions of all concerned should be regulated. Owing to the fact that many of the employees are women and girls, men should be chosen to fill executive posts, that some chance be given to their weaker colleagues to make their escape, whatever occurs. Upon the first signal of the alarm, each member of the staff should, as in factory drill, clear all gangways of either rubbish, stock or obstructions. They should then form in double lines along the aisles leading to the exits. Those not actively employed in the emergency organization should then form squads and in pairs, women holding up their skirts, march to the exit they have been previously instructed to use. The elder women in all departments should be trained to lead these lines and, incidentally, their example is sure to have a steadying effect upon both their subordinates and their customers. But the real problem is concentrated in how effectually to deal with the casual public, who throng these buildings daily to so great an extent that it is estimated not less than ten thousand persons are sometimes on the premises at the same moment. Apart altogether from the private fire brigades maintained by these establishments, apart altogether from fire escapes and the most modern fire precautions, this constitutes the real peril which must be initially overcome. It can only be accomplished by constant and painstaking training of every individual employee and by their example, coupled with the exertions of aisle guards, who will indicate to the flurried and hysterical, how safety may be most easily reached. At such a moment one cool floor-walker with his wits about

him will potentially save more lives than the best equipped
fire department which ever travelled the streets.   Conquering a fire is one thing, conquering a panic another, and,
whereas even after its inception, prompt action may quell
the former, the latter belongs to the elemental side of human
nature and as such is beyond the reach of science or apparatus.

As for structural safeguards, the disposal of rubbish
and other means towards fire control, they are dealt with
elsewhere, though the cardinal factor of human temperament can never be altered or modified by such external measures, except inasmuch as the knowledge of
their presence tends, to a certain degree, to alleviate
fear.

If the employees of a department store are faced with
an enigma in dealing with their customers, then most assuredly the staff of any hospital have every reason to fear
fire and its accompanying risks.   Here the problem is complicated by the absolute helplessness of the patients and the
possibility that severe shock, in some ·instances, may result
in death.   But on the other hand, nurses, attendants and
doctors are all persons of superior intelligence and may
be expected to carry out instructions, not like automatons,
according to the letter, but with due regard to prevailing
conditions.

Equally, fire control in hospitals has for long absorbed
the ingenuity of architects with the result that, generally
speaking, they are well safeguarded.   At the same time,
however, certain simple devices can easily be installed,
amongst the most valuable of which are "fire breaks,"
which acting automatically accomplish in corridors with
flames precisely what watertight doors accomplish in ships
with water.   In other words, they delay the enemy, and
if unsuccessful in their passive defense, at least hold him
in check long enough to insure the adoption of precautionary measures for those concerned.   In simple language,
they may be described as iron drop-doors, which being
operated, cut off the area involved from the rest of the

building. Further, though properly speaking this is a structural safeguard, the employment of fire-towers is strongly to be recommended. These consist of a covered staircase adjacent to, but distinct from, the main building, and connected by iron gangways at each floor. Thus, by closing the exit doors, which are fireproof, a completely isolated staircase is formed, down which patients can be moved to safety without hurry or alarm. Incidentally, these towers form admirable adjuncts to all classes of structures and public edifices habitually frequented by numbers of persons of both sexes and all ages. It seems unnecessary to insist once more upon the careful disposal of all rubbish, since such a precaution appears to belong to the obvious. But the writer's experience has taught him that it is precisely the most ordinary safeguards which are habitually neglected.

Finally, the following recommendations may be accepted as applicable to all classes of buildings, and if adopted, promise a large measure of safety for occupants.

(a) All stairways or a sufficient number of them should be located in fireproof shafts, having no communication with the main structure, except indirectly by way of an open-air balcony, or vestibule, on each floor. Hose connections attached to standpipes should be located on each floor in the stair-towers available either for public or private fire department use.

(b) Stairs, if any, inside the building, and elevators, should be enclosed in shafts of masonry, and have fire doors at all floor communications.

(c) Old buildings with inadequate fire escapes should be provided with automatic sprinklers, or smoke-proof stair towers, but outside fire-escapes passing in front of or near windows should be discouraged.

(d) All factory buildings employing operators in the manufacture of inflammable goods should be fitted with automatic sprinklers, and this system should likewise be extended to all classes of structures generally frequented by a considerable number of persons.

(e) Large floor areas should be subdivided by fireproof partitions or brick walls.

The above are not counsels of perfection and are well within the reach of those having the safety of their fellow-creatures at heart.

CHAPTER XVI

FIRE-FIGHTING IN THE UNITED KINGDOM

COMPARISONS are often instituted between the fire risks of London and New York. It is glibly pointed out by statisticians with facile pens that, whereas for every one hundred thousand inhabitants, London averages eighty-one and New York three hundred fires, and whereas the population of London is considerably greater than that of New York, *ergo,* fire control in the former city has attained to a higher degree of scientific evolution than in the latter, and further deductions are drawn according to the nationality and enthusiasm of the individual. But such reasoning is founded upon the superficial aspect of the subject, without taking into consideration the numerous contributive factors governing the problem. In the first place, in making these invidious comparisons, writers forget that units of apparatus, ability of personnel and general efficiency may be on a par in two separate fire-fighting organizations, but, owing to local causes and climatic conditions, the annual record of the two may be widely divergent. New York has certain disadvantages with which to contend. Its modern buildings are the highest in the world, in themselves a staggering question for the fire-fighter; in portions of the city there are streets comprising nothing but wooden buildings which burn like torches; extremes of climate render fires more prevalent and more hazardous; the alien population is vast and criminally careless; and, finally, unfortunately, "arson" has grown to be regarded amongst undesirables as a legitimate and easy way of obtaining insurance as a form of income. Any unprejudiced observer will allow that in these respects London is more fortunately situated, which admission detracts in no whit from the standard of excellence of both departments, which are well worthy of the great capitals they represent. In considering rather more particularly the outstanding features of its brigade, London, in common with all other English cities, has no need for the heavy appliances usually seen in America. Owing to the narrow tortuous streets in all ancient towns, the manipulation of weighty apparatus with lengthy wheel bases becomes practically impossible, while the average building constructed of stone and of only four or five stories in height does not constitute a grave fire risk. The residential area of London is chiefly composed of such erections, and even in the wealthier suburbs, where the houses stand in their own grounds, this fact of itself is sufficient to prevent a serious conflagration. In the eastern section of the Metropolitan district and along the docks and wharves lining the river Thames, the risks are materially greater, and hence there is a considerable concentration of strength in this locality, though it is worthy of note that of the sixty-five outbreaks classified as serious during the year 1911, about thirty-four occurred in quarters inhabited by aliens, a sufficiently good indication of the truth of the preceding statement, that these foreign colonies are a fruitful source of anxiety to the authorities of any city. It is also of interest that of the grand total of four thousand and odd fires during that year, no less than 1762 were directly attributed to carelessness, while only twenty-eight resulted from arson. Prior to the year 1866 the protection of London from fire depended upon an organization known as "The London Fire Engine Establishment," which consisted of only one hundred and thirty officers and men, operating seventeen stations. The cost of its maintenance was chiefly borne by fire insurance companies and its duties were practically confined to "fire quenching." For the saving of life from fire during many years, the "Royal Society for the Protection of Life from Fire," which was supported by voluntary contributions, supplied and manned some eighty-five fire-escapes which were stationed in various parts of the city, few being in the suburbs.   There were in addition, several so-called volunteer fire brigades, which were not under the direct· control of any recognized authority, and proved a veritable thorn in the flesh to the Municipality.   Having no definite financial support, they employed "collectors," with the result that the unfortunate rate-payers were solicited for contributions towards associations which, with rare exceptions, seldom performed any useful service, and for the disposal of whose funds there was no adequate guarantee.   Moreover, it was customary for these gentlemen to wear a uniform similar to that of the professional fire-fighters, causing a good deal of acrimonious confusion amongst those who were under the impression that they had contributed towards the funds of the professional fire brigade.

The Act of Parliament of 1865, however, put an end to this ambiguous state of affairs, and on January 1st of the following year the Metropolitan Board of Works assumed the responsibilities of the two first mentioned organizations, under the title of "The Metropolitan Fire Brigade."   As for the volunteers, they lingered for some time, being disbanded one by one, till in 1900, the sole surviving company, located in Islington, closed its doors. The funds provided by the aforesaid Act for the maintenance of the Brigade were: (A) contributions by the fire insurance companies at the rate of £35 ($175) per million of the gross amounts insured by them in respect of property in London.   This was calculated to bring in about £10,000 ($50,000) annually, while in addition, the buildings and staff of the "London Fire Engines Establishment" were handed over free of charge. (B) A government grant of £10,000 ($50,000) a year, in consideration of the protection afforded public buildings and offices. (C) The produce of a halfpenny rate (1 cent) on all the rateable property in London, which, it was estimated, would realize a sum of £30,000 ($150,000) a year.   It will thus be seen that less than half a century ago it was decided after careful debate that the fire control of London could be accomplished for the expenditure of £50,000 ($250,000) per annum, whereas today it costs in the neighbourhood of £270,000 ($1,350,000) or more than five times as much, and there is no probability of any finality having been reached.   To defray this constant increase, varied legislation was introduced, till the Local Government Act of 1888 virtually repealed any limitation of the amount which might be raised from the rate-payers for fire brigade purposes. Incidentally, it is an interesting historical fact that the year 1865, which saw the birth of the modern London Fire Brigade likewise witnessed the genesis of the existing New York Fire Department.

The first Chief of the newly formed organization was the late Captain Sir Eyre Massey Shaw, K.C.B., who, an Irishman by birth, had previously been in charge of the Belfast Fire Department and whose subsequent twenty-five years of service with the London command witnessed the stations of the Brigade quadrupled, and the strength of the personnel increased from one hundred and thirty to seven hundred men.

Amongst the more important changes which he introduced were the street fire alarm system and the substitution of telephonic for telegraphic communication between stations.   From its formation until 1904, the force was known as the Metropolitan Fire Brigade, when, with the sanction of Parliament, it was designated the London Fire Brigade. The old title was somewhat misleading, since large districts in the London area are outside London county proper, and though on occasion the services of the Brigade may be summoned to assist suburban fire departments, it is then entitled to make a pecuniary charge for the same.   Thus, for the attendance of a steam fire-engine the scale of payment is a preliminary fee of £2 ($10), with an additional £1 ($5) for every hour or part of an hour during which it may be working.   A fireboat costs as much as £6 ($30) for the initial expense of its attendance, each succeeding hour being rated at £1($5).   The manner in which this expense must be borne is clearly indicated in the following excerpt from the official regulations: "In such cases the owner and occupant of the property on which the fire occurs are jointly and severally liable to pay a reasonable charge in respect of the attendance of the brigade." This in itself is sufficiently clear, at least in theory, but in practice there must at times be some difficulty in collecting such charges, in which case presumably the ratepayers of the borough concerned must be held responsible. Without wishing to be hypercritical, it does appear to the writer that such a system is open to grave disadvantages, since it is seldom the destruction of one individual building which is at stake, but rather the possibility of the fire spreading and endangering a large area. On the other hand, it may be argued that those who do not assist in supporting the organization have no right to expect the free use of its apparatus and personnel. Hence, it becomes a question beyond the criticism of one not conversant with local conditions. To accommodate the London Brigade there are eighty-three stations or engine houses, though in addition there still remain some of the old "street stations," wooden shelters in which as a rule are kept extension ladders and which, owing to the inconvenience they cause vehicular traffic, are fast being superseded.

For the extinction of riverside or wharf outbreaks, fireboats are stationed at certain points on the river, their crews being lodged either in adjacent fire-stations or in buildings especially erected for the purpose. The general principle determining the distribution of fire-stations in London is the necessity of ensuring (a) the speedy arrival after a call of life-saving and fire-extinguishing appliances at any spot in the protected area; (b) the concentration of one hundred men within fifteen minutes in any dangerous location for large fires. On receipt of an alarm, the fire appliances turn out with all possible speed, sliding poles for men on duty and automatic hitches for harnessing the horses being now of almost universal adoption and rendering a start feasible within the space of a few seconds. It is the custom for the life-saving appliances to leave first, which on the face of it is humanitarian, though perhaps not eminently practical. Unaided by other apparatus, except under rare circumstances, it can accomplish little. All appliances at the station nearest to the scene of an outbreak are withdrawn for service, leaving one man behind in the watch-room to preserve telephonic communication between the officer in charge at the fire and the district superintendent, while the former has power also to obtain the assistance of the engines located at the station next to his. Arrived at the outbreak, the fire chief classifies it according to its severity, and transmits one of the three following calls: 1.—The "Home call," which signifies that he is confident that he can deal with it by means of the apparatus at his command. 2.—The "District call," which means that more assistance is required, but that it can be obtained from adjacent stations, and, 3.—The "Brigade call," which is an appeal to headquarters for both men and apparatus. Such a system, certainly, as applied to any city with a considerable fire risk, is open to grave defects. In the first place the telephone operator left at the station of the first alarm is in a difficult position if another alarm comes in from the surrounding neighbourhood, and might well become flurried and misunderstand orders. It is imperative that there should be direct communication between the Fire Chief and his lieutenants, otherwise there can be little hope of effective coöperation. In addition no provision is made for one engine house to cover another, so that in the event of several fires in the same area all can be quickly and efficiently attacked. The policy employed appears rather to be one of centralization of men and apparatus, robbed of half its efficacy by roundabout and clumsy means of giving and receiving orders. At any rate, it is certain that in a city like New York the practice of taking all the appliances from a fire station and leaving it unprotected would be fraught with the most terrible danger to public property and would, in fact, never be tolerated by those responsible. Local conditions may vary and materially influence fire organization, but the cardinal points of fighting strategy are the same the world over and, as in regular warfare, one of the most important is ever to be prepared for a flank or rear attack.

The present strength of the London brigade consists of one Chief Officer, four Divisional officers, 211 subordinate officers, and 1,163 rank and file, including men under instructions, pilots and coachmen. Considering the prevailing wages in England, the scale of payments is adequate, starting with £1,000 ($5,000) a year for the Chief, a maximum of £245 ($1,225) for Superintendents, while the ordinary fireman receives 35/- ($8.75) a week after qualifying as efficient and passing certain tests. Quarters are provided for single and married men alike, a minimum charge being made for the same ranging from 4/- ($1) per week for married firemen to 1/- (25 cents) for single men. Incidentally it is compulsory to live in these quarters, which are in every case situated over the engine houses. After five years of approved service and less than fifteen, there is a gratuity of one month's pay for each year of such service. Upon the completion of fifteen years, three-tenths of the pay then being received is allocated as pension. Those who serve for a longer period receive corresponding increases, until, with twenty-eight years of service, the pension amounts to two-thirds of the nominal pay. In the event of those incapacitated from further service through injuries received in the execution of their duty, a special allowance is made according to the particular merits of the case. Pensions are also allowed to the widow, and children of officers and men killed while on duty. The regulations are unusually considerate in that in the case of a widow remarrying, although her pension is to be suspended from the date of her remarriage, should she for the second time become a widow, it may be restored on proof that her circumstances are such that it is necessary for her support and that she is deserving of the public bounty.

For many years it was customary to enroll as firemen

none but seafaring men, but lately this has been modified and now entry to the brigade's ranks is open to all possessed of the necessary qualifications. As regards promotion, this is limited, inasmuch as the senior officers of the force are usually drawn from the executive branch of the Royal Navy, or are engineers of repute in some specialized section of their chosen profession. This forms a radical difference between the practice, not alone of the United States, but of most foreign countries. True, in Germany and France, senior officers are men of naval or military training, but they join the fire service as youths and work their way up through the various degrees of command in a precisely similar manner as they would were they attached to the Army or Navy. But the point is that they are thoroughly trained at every step and though the system is by no means ideal, it is preferable to the appointment of officers to a highly scientific corps, who though no doubt able and intelligent men cannot possibly possess either theoretical or practical knowledge of the subject with which they are called upon to deal. In America it is no exaggeration to say that every newly enlisted fireman is a potential chief; it depends solely upon the ability and determination of the individual whether or no he shall rise to a position of executive importance, or whether he shall spend all his days in the ranks. This must prove a powerful incentive to any normal character to go forward and win, and it may be asserted without fear of contradiction that the fireman devoid of ambition is of little use to any fire department. It has been argued time and again that the rank and file are apt to prove insubordinate when one of their number is delegated to be in command over them. Certainly in American fire practice this has never proved the case. Rather has it been the opposite, and the men have been proud in the success of one of their comrades. Further, in the opinion of the writer, the profession of fire-fighting is one which demands that those adopting it as a calling should be equally versed in all branches of its requirements in practice and theory. This entails actual experience in physically as well as mentally fighting fires. It means familiarity with the handling of hose, the management of extension ladders and that intimacy with fire as an enemy which can only be gained first-hand. It goes without saying that the officer in charge of a battleship's barbette guns is thoroughly conversant with every detail of their mechanism, with their muzzle velocities, with their arc of fire, with the individual merits of their projectiles, be they armour-piercing or high explosive. In fact, he is a master of craft learned in the school of experience. Why then should less be expected of, or demanded from, officers in charge of equipment with which to fight every whit as dangerous an enemy as ever sailed a sea, when lives unnumbered and property beyond calculation may depend upon their actions, and when it must be remembered that every fight is *a l'outrance,* without chance of armistice.

Turning now to the matter of appliances, the eighty-six fire stations are equipped with seventy-two horsed steam fire-engines, three steam and fifteen petrol motor fire-engines, sixteen mechanically driven fire-escapes, one hundred and ninety-one ordinary fire-escapes, ninety-four hook and ladder trucks, eleven hose and coal vans, one motor lorry, ninety hose carts, and fifty-six miles of hose, not including a large amount of smaller apparatus. Of the steam fire-engines, it is noticeable that some of them, though of antiquated pattern, are still adequate for useful service, thus one has been in use since 1878, or over a quarter of a century. All the modern types are double-cylindered, their average pumping capacity being one and three-quarters English gallons per revolution. In every case, axles and wheels are made to gauge and are interchangeable. Some of the engines are fitted with burners for using petroleum as fuel. For a number of years it has been the practice to keep a sufficient pressure of steam in the boilers of all engines to enable a full working head to be obtained in from two to three minutes after leaving the station.

Since 1901, motor traction has been gradually introduced, and though the steam-propelled fire-engines have not given entire satisfaction and are even now for sale, there is no doubt, as elsewhere, that motor traction has come to stay, and in course of time the horse will be eliminated from the department. As regards ladders, the largest in use by the brigade are those of eighty-two feet, that being their vertical height from the ground when fully extended. Being heavy and requiring a strong crew, they are used primarily for facilitating fire-extinguishing operations. The work of their extension is controlled by a small motor, worked by a compressed carbonic acid arrangement, or in emergency by hand, and being mounted on a turn-table fixed to a horse-drawn carriage, are known generally as "turn-table long ladders." For regular "fire-escape work," the ladders used are fifty-five and seventy feet, of the telescopic pattern. Motor-escapes are being introduced, but their use has by no means become general. As regards horse-drawn appliances, which are still largely in the majority, the animals are hired from job-masters, and the price paid, including ordinary harness, fodder, straw, and stable utensils, amounts to about £70 ($350) per horse per annum. The job-masters take all risks. At stations where automatic harness is in use belonging to the department, a reduction of 55/- ($13) is made. Practically all the horsed escapes are fitted with appliances known as first-aid fire-extinguishing machines. These consist of a tank containing water connected with a cylinder of compressed air, which, being operated, can maintain a jet for from four to four and a half minutes. There are about eighty-one of these appliances in use, but the chemical engine, *per se,* is unknown. Such additional apparatus as cellar pipes, smoke helmets and a small number of hook ladders are in regular use, but it is noticeable that the water tower and other forms of heavy equipment are lacking, which is to be accounted for by the low buildings and narrow streets. It is almost a pity that a picturesque survival of ancient days has latterly passed from the London fire department. Not many years ago it was customary for firemen, proceeding to the scene of outbreak, to herald their progress by shouting

"Hi Hi Hi!" Some five years ago this was discontinued, and a brass bell substituted.

Some reference may here be made to the floating fire equipment of the London brigade. This has been developed under peculiarly restricted conditions, since the Thames is a tidal river with only five feet of water in places at low tide. Moreover, most of the traffic is carried on in heavily laden barges, with low free-boards, handicapping fire boats as regards their speed, owing to the liability of their wash swamping the .former. Finally, these "fire floats" are limited as to their length and breadth, since it is necessary so to arrange their dimensions that they may easily enter the connecting locks of docks and canals, while their height above the water line must also permit them to pass under low bridges. Considering the enormous dock area of the Port of London, the greatest credit is due to the officers and men for the ability and technical skill which they display in defending the vast responsibility committed to their charge, which, owing to the conditions already stated, rendered the problem one of the most difficult of solution in the world. The first vessel to be built was the "Alpha," a twin screw boat, eighty feet in length, with a beam of sixteen feet and a draught of three, on a displacement of sixty-three tons, and having a speed of ten knots. It was fitted with pumps capable. of discharging 1,250 English gallons a minute. This was succeeded by another handy little vessel, the "Beta," equipped with four pumps, each with a discharging capacity of 1,000 English gallons a minute at 140 pounds pump pressure. These have been followed by the introduction of motor fire floats, propelled by internal combustion gasoline engines. The latest addition embodies several new principles. With a length of one hundred feet and a beam of nineteen, the draught of water is only two feet. She is propelled by triple screws driven by sixty horse power engines, and the collective delivery of her pumps is about 1,500 English gallons a minute. The screws work in tunnels, and her design is such that when proceeding at full speed, about eleven knots, the bow wave is absorbed under the bottom of the vessel and little or no wash is apparent.

With the object of ensuring that the fire brigade shall be readily and easily summoned when a fire occurs, great attention is being paid to increasing the facilities by which it may be called, and in making them known to the public. Connected with every engine house are fire alarms situated in the chief thoroughfares, the number varying with the fire risks of the area. These alarms are equipped and supervised by the Post Office, an annual rent being paid for their use, as also in the case of the station telephones. In 1911, the number of these alarm boxes totalled 1,545, and the Council of the fire department have a scheme for their augmentation on a large scale at present under consideration. It is worthy of note that a number of posts in the East end of the city have recently been fitted with tablets bearing instructions for their use in Yiddish. All street alarms are adapted for the transmission of telephone messages by firemen, suitable instruments being carried on all brigade appliances for this purpose. The principal police stations are in telephonic communication with the fire stations, and this also exists between the latter and the majority of public and other large buildings, such as theatres and so forth. During 1911 the total number of calls amounted to 6,868, working out at a daily average of nineteen, through on the 14th of August, during a great heat wave, there were fifty-three. In England it is a criminal offense to send in a malicious false alarm, and a person so doing is liable to a penalty not exceeding £25 ($125) or imprisonment for three months with hard labour. From the latest report it appears that the total of malicious false alarms has increased considerably, amounting now to 357, which constitutes a record. This increase can only be explained by the unwelcome attention of the suffragettes with militant tendencies. It is outside the scope of this work to enter upon a dissertation concerning the rights and wrongs of the movement, but as far as fire duty is concerned there can be only one opinion. Every false alarm draws men and apparatus away from a certain area, which is thus left so much the less prepared to meet an attack by fire, while it stands to reason that constantly responding to these calls unnecessarily fatigues men and horses and renders them less fit for duty. Hence, the action of these women constitutes a menace to the community and is one of selfish egotism deserving not alone the condemnation of every right-minded citizen, but the infliction of such punishment as shall render similar behaviour in the future, even on the part of half-crazed fanatics, unlikely to occur. There is a considerable fluctuation in the number of fires reported annually in London. Thus, in 1907, the total of 3,320 was a decrease of 523 on the preceding year and represented a financial loss of £493,389. 1908 and 1909 saw the decrease continue, the latter year having 123 fires less than 1907, though the financial loss shot up to £699,329. It remained, however, for 1911 to beat all previous records, the number of fires amounting to 4,403, an increase of 1,195, monetary loss being £789,003, or nearly four million dollars. The extended use of motor vehicles has an important bearing both on the outbreaks registered and on the amount of the fire loss, since fires in garages are common and the values involved considerable. Chimney fires are not included in these statistics, though their number averages about 800 per annum. In this connection it may be remarked that, by a special Act of 1900, the occupant of any house the chimney of which catches fire, must pay towards the cost of the London fire brigade, a fine fixed upon the rateable value of the premises ranging from 2/6 (.60), up to a maximum of £1 ($5).

The training of the men belonging to the brigade is sufficient for the demands made upon them, particular attention being given to motor engineering, instruction in first aid, and gymnastics. The following excerpt, however, taken from the last official report issued, throws a curious light upon the somewhat haphazard methods employed in the physical training of the men: "Early in 1911 the fitting up at the Manchester Square fire station of a gymnasium

with apparatus for carrying out Swedish drill was completed
and an additional fireman was added to the strength of the
Brigade to act as physical drill instructor. So much interest was evinced in the matter by the staff, that gymnastic
apparatus has been provided at other stations and a number
of well attended classes has been held. There is no doubt
that the staff have felt the lack of opportunities for physical
training, especially in view of the number of hook ladders
carried on fire appliances and the extended use of such
ladders."

It seems needless to emphasize the importance of constant physical training of firemen, as, quite irrespective of
a man's muscular development, he quickly becomes stiff and
slow of movement unless either constantly drilled or at least
given the opportunity of obtaining gymnastic exercise. As
regards loss of life, during the last year for which particulars are available, 151 deaths are recorded, of which no
fewer than sixty-two were those of children under the age
of twelve. The causes of the fires. at which this loss
occurred include twelve cases of children playing with fire,
eleven from their playing with matches, while thirty-two
were attributable to clothing coming into contact with fire
or gas stoves. This infantile mortality of approximately
forty per cent. of the total death roll, though in itself insignificant when compared with an estimated population
of over four and one-half millions, certainly points to the
fact that some sort of instruction anent the dangers of
fire could usefully be included in the curriculum of Board
Schools, as is done in Germany.

By the Metropolis Water Act of 1871, it was provided
that the water companies should supply, where necessary,
water for fire-fighting free of charge, while they should
also install such plugs or hydrants as might be required
at the expense of the department. At present the total
number of hydrants in the Metropolitan area is about
29,000. Before 1897 all hydrants in the County of London,
outside the City proper, were made with one outlet. In that
year it was suggested that new hydrants should be provided

with two outlets when erected in localities where the fire risks were considerable. The disadvantage of the double outlet hydrants used in the City itself, numbering over eight hundred and provided by the City Corporation at its own cost, was, that in the event of it being necessary to connect a second length of hose to the hydrant when the first was already in use, the control valve had to be temporarily closed. Apart from delay, which might be fatal, the fireman operating the branch in action was often disconcerted by the stoppage of his water supply and might conceivably find himself in danger owing to this cause. To obviate this the Fire Committee of the London County Council decided that the experiment should be tried of placing two hydrants in one pit, each being fitted with a control valve. This scheme proved satisfactory, though, on account of the expense involved, its introduction has not been general. In January, 1901, however, it was determined that the branch pipes connecting hydrants to mains should in every case be of five inches diameter, i. e., sufficient to supply double hydrants should they be universally installed. Hydrants fixed in public thoroughfares are tested by firemen every two months, and are also examined and tested by inspectors under the supervision of the Chief Officer, and as a result of deficiency in water supply on some occasions many fresh connections have been made, and the water companies involved have themselves contributed towards the expense of laying new mains or pipes for fire purposes. The quantity of water used during 1911, amounted to thirty-three million odd gallons, or not three times as much as was used in one fire in New York—the Equitable. Two-thirds of this quantity was drawn direct from street mains, the other third being supplied by the river Thames and canals. The double pattern hydrants deliver on an average 800 English gallons a minute which does not seem excessive when compared with the 4,000 gallons obtainable from the "high pressure" hydrants in America.

The inspection of public buildings is undertaken by the Brigade free of charge, though in certain cases grants in aid are forthcoming to provide for the special staff necessary for the operation. In theatres, in which public performances are regularly given, there is an official inspection every ten days to ensure that the rules of the Municipality for securing the safety of the public are enforced. This applies, in a modified form, to cinematograph halls, temporary exhibitions and bazaars, and the plans of all new buildings requiring licenses, and of proposed alterations to existing buildings, are referred to the Chief Officer for examination and report.

Lodging houses designed to accommodate more than eighty persons are likewise under the control of the fire department, as regards means of escape, etc., and though in the first instance this is the business of the architect who must conform to the building rules of the London County Council, the responsibility for their efficient maintenance rests with the fire chief. In all, about 22,000 inspections are made annually. Latterly, the underground electric railways, which honeycomb London, have also passed under the supervision of the Chief Officer whose advice has been followed regarding the fire safety of these means of transportation. It is estimated that the cost of this inspection branch amounts to approximately £4,000 ($20,000) per annum. No less than two hundred and sixty-five officers and men were injured in the execution of their duty during the year 1911, or one-fifth of the whole fire fighting strength. This is a high percentage and bespeaks devotion to service which is in every way commendable, and it must be remembered that it is not always the greatest fire which offers most risk to life and limb. Thus it will be seen that London is efficiently guarded in its fire hazard and though, perhaps some of the methods employed may appear antiquated and not in accordance with the latest improvements in fire control, yet after all "the proof of the pudding is in the eating," and for a city of its population the fire loss is small. Whether or no the department could successfully cope with a great sweeping conflagration in the warehouse district is a moot question, which most assuredly the writer trusts will never arise for solution. But there seems no doubt that the chances of a serious disaster from fire in the residential district are practically non-existent, while the building regulations in force are of so stringent a character, that except in the case of panic, against which no one can guard, fire risks in theatres are reduced to a minimum. In this connection a word of praise must be added for the British Fire Prevention Committee, a voluntary organization which has devoted time, energy and money towards the solution of all problems affecting fire control, and which, in the most public-spirited manner, has given the results arrived at "gratis" to the world at large.

In considering the fire departments of the important Boroughs in the United Kingdom, one outstanding feature is little short of amazing. This is the small number of the personnel, as compared with the population and the pecuniary values they are called upon to protect. Of course, the fire risks are appreciably smaller in residential districts than they would be in America, since houses are commonly constructed of stone, are of limited height, and generally do not offer themselves an easy prey to the flames. But far otherwise must it be in the congested warehouse section, and it is really marvelous that disastrous fires are not of more frequent occurrence. It will be of interest to describe *seriatim* the brigades of four great provincial cities, Belfast, Birmingham, Glasgow and Manchester.

Belfast, the commercial capital of Ireland, is a city of 385,000 inhabitants. During the year 1911, the number of fires amounted to 228, with an estimated fire risk of nearly one million pounds sterling, the actual loss amounting to fifty two thousand odd pounds (two hundred and sixty thousand odd dollars). Yet the strength of the Brigade consists only of seventy-five firemen, including the superintendent, assistant superintendent and third officer. The plant includes sixteen fire engines, three petrol driven escapes, one eighty-foot extension ladder and various other smaller apparatus, all of which seem to be absolutely up to date. Though it may be remarked that salvage work is accomplished by a petrol salvage motor trap, yet in addition, there are two horse and one motor ambulances, which responded to over three thousand calls in that year, covering a distance of 10,194 miles and occupying thirty-two minutes per journey, i. e., from the receipt of the call until their return to the station. These ambulances are used for ordinary accidents. The steam fire engines were only used once, and the motor fire pump only twice during the entire year. Machines traveling to and from fires averaged eight miles for each "turn-out," and were engaged for 133½ hours, or an average of thirty-five minutes for each fire, this calculation including the journey to and from the outbreak. These are remarkable figures, and since, within the last ten years, there has been practically no increase in the number of fires, Belfast can congratulate itself upon having one of the most economical and effective systems of fire control, salvage work and ambulance equipment probably in the entire world. Birmingham, according to the last census, is a city with a population of 840,200, covering an area of 43,500 acres. The fire department consists of a Chief Officer, two senior subordinates and 194 rank and file, which represents roughly one fireman to every 221 acres or to 4,200 persons. One thousand and forty-eight alarms were received during 1912, of which one hundred and eight related to chimneys on fire, one hundred and twenty-six were false and six were "malicious." The estimated value of the property at risk was over three and a half million pounds, the actual loss approximating £81,000 ($405,000). On an average, these alarms occupied only thirty-seven minutes each, while on twenty-eight days no calls were received, on sixty-seven days only one, on seventy-five days two, on twenty-four days six, on two days eight, and on one day eleven, these with some forty stations and nine thousand hydrants. The police department house some of the apparatus in their quarters. Amongst the most important appliances may be noticed six motor turbine pumps and escapes, twelve steam fire engines, one water tower—

a unique feature in English fire practice—three chemical engines and twenty-one extension ladders. Like Belfast, there is also an ambulance corps, manning no less than eleven ambulances, while amongst minor apparatus may be noticed nine smoke helmets of the latest oxygen battery type. It is worthy of note that every man in the Brigade possesses a first-aid certificate for ambulance work. The Chief Officer of the department receives £400 ($2,000) per annum, while ordinary firemen are paid from 24/- ($6) to 31/- ($7.75) per week. All ranks receive free quarters, light and uniform, eight pence per week boot allowance and six pence (twelve cents) washing. Annual leave is granted to the extent of eight days for a fireman, with an addition of sixty hours a month taken in two periods of twenty-four hours and two of six. All places of amusement are inspected by the department, while public and other buildings under the direct supervision of the Brigade, as regards fire risk, pay a small annual charge, the proceeds of which are devoted to the recreation and superannuation funds. This again constitutes a remarkable record for a small, though excellently equipped department.

Glasgow, which since the commencement of 1912, includes Govan and Partick, has a Brigade the authorized strength of which is 195 and is at present fifteen short of this number. During 1912, engines and firemen operated at 526 fires, while in 110 cases the outbreaks were so trifling that they were suppressed with hand pumps. The estimated loss amounted to £150,000 ($750,000). It is noteworthy that fires reported as due to defective building construction amounted to 202, or over thirty-one per cent. There are eleven stations, housing one motor extension ladder, sixteen motor pumps, eight steam fire engines, and two motor first-aid traps. In the entire department there are only two horses, which is a sufficient indication that with true Scottish acumen motor propelled vehicles have been found cheaper and more effective. *Sic transit gloria equi.* Four first-aid motor machines are in course of construction, each being designed to carry one officer, twelve men, two thousand yards of hose, one thirty-foot extension ladder, an ambulance box, tools, and other necessary gear. When fully laden, these motors will weigh about fifty-five hundred pounds, with a length of nineteen feet on an eleven-foot wheelbase. The number of malicious alarms was peculiarly high, amounting to no less than fifty-nine, with only six convictions for the same. Attendance at fires under certain conditions must be paid for, and the income from this source amounted to nearly £4,000 ($20,000), while listed amongst "special services rendered" are the two following interesting items; "entering houses for locked-out tenants," 120 occasions, and "searching the roof of a building for thieves," once. Without wishing in the least to be ribald, the writer cannot help wondering why the duty of assisting burgesses of Glasgow, who had either forgotten their latch-keys or, perhaps—such things do happen—had been locked out on purpose, should have been delegated to members of the fire brigade!! The worst outbreak of the year was that caused through the ignition of a cinematograph film, while in process of manufacture. The fire was under control within half an hour, but not before damage to the extent of £5,000 ($25,000) had been done. This led to an inquiry into the whole subject and it was found that in one establishment the basement of which was heavily stocked with this inflammable material, the upper stories immediately above were utilized as an hotel. It is hardly necessary to dilate upon what would have occurred to the guests had the fire broken out. This incident is mentioned since, no matter how well a building may be constructed, danger of this sort cannot be invited with impunity, especially when the personal safety and property of 785,000 people rest upon 180 firemen.

The report of the Manchester Fire Brigade is again remarkable for its brevity and for the fact that the authorized strength of the force is only 130, including officers, for a city of 715,000 inhabitants. All the world knows that within this area are to be found some of the greatest cotton spinning factories extant, and to the outsider it certainly would appear as though the fire-fighting force could not be adequate for possible demands. For instance, supposing there were three outbreaks of even moderate size in different parts of the city at the same time, a perfectly normal contingency to contemplate, how could they be successfully attacked? The 524 fires of the year 1912, represented a property value of over three million pounds (fifteen million dollars), though the loss was only £102,000 ($510,000). The firemen were actively engaged during the entire year for 320 hours, 35 minutes, this including false alarms, or under an hour a day; while the fireboat responded to seven calls, and was actively engaged on only three occasions. Now, it may be argued that the four men forming the crew of the fireboat, or the 112 men rated as "firemen" in the Brigade, earn their pay with an absence of worry or anxiety which might be envied by the layman; in fact, doubtless the rate-payer reads the report in question and contemplates the pay-roll dubiously, revolving the while in his mind whether the total expenditure is justified, or whether, after all, it is not a piece of gross municipal extravagance. The answer is no difficult one to give. When a man insures his life he pays a premium for certain benefits of which, perhaps, he may never taste, but on that account he does not cease his payments. Similarly, with all outlay for all contingencies, there is no direct and immediate return that can be touched, handled and assessed at so much material value. But none the less the value is existent though not perhaps to the extent demanded by a captious rate-payer. It is, in short, a payment for municipal fire insurance, and though day after day and month after month the protected area may jog along with no serious outbreak to trouble the even tenor of its way, the time may come when every man and every piece of apparatus will be engaged in a life and death struggle for mastery. And it is precisely against that event that the municipality, which is far sighted, guards. Hence it is that with the greatest of deference to those concerned it does strike the writer with something akin to amazement that such colossal values should be so lightly guarded as they apparently are in English provincial towns. For, given the best of appliances and most skilled firemen, what could 130 men accomplish against anything in the nature of a sweeping conflagration? And supposing other fires occurred at the same time, it would be a physical impossibility adequately to proteet against the one or attack and quell the other.

As an example, it is not necessary to travel beyond the British Isles. In August, 1911, there occurred a serious fire in the Carlton Hotel, London, a building of moderate size and certainly of no greater magnitude than some of the hotels to be found in the towns, mention of which has been made. This outbreak necessitated the employment of twenty-three steam and motor fire-engines and the attendance of two hundred and two officers and men before it could be brought under control. Had this occurred in Manchester,—well it is needless further to comment!!!

This is penned in no carping spirit and with the knowledge that man for man English fire departments are the equal of any in the world, but they cannot accomplish miracles, and rather are the municipalities to blame who, secure in the traditions of the past and unmindful of the chances of the future, are so penny wise and pound foolish that they are ready to risk millions of pounds worth of property in order to escape an infinitesimal addition to their rates. Place the whole question on a business basis, work out the value of the fire insurance premiums paid on the property within the municipal area and compare the total arrived at with the total expenditure per annum on the fire departments under discussion and the result will perhaps surprise owners and insurance companies alike. Of course, it may be argued, that it belongs to the business of the latter to assess their own risks and avoid the acceptance of policies in badly protected areas. But that is outside the main discussion, which is concerned with the ethics of fire fighting. And most assuredly he would be a bold man who would prophesy that fire would never conquer under such conditions.

CHAPTER XVII

THE NEW YORK FIRE DEPARTMENT

AN experience of a quarter of a century as an active fire-fighter has left one indelible impression upon the brain of the writer. It has been his good fortune to meet professional colleagues hailing from every known part of the globe, while, equally, he has had abundant opportunity to inspect the fire departments of the great American cities as well as those maintained by European municipalities. And one point in common he has found them all to possess; namely, that they are firmly convinced that they belong to and represent the latest, the greatest and the most up-to-date fire department in the world. Which is to be expected; if a mother be not proud of her offspring, who should be?

And that very enthusiasm itself speaks well of the calling as a whole, and is a sufficient proof that its votaries bring to bear upon their occupation all that interest which is necessary to make of a chosen life work, a success. Hence there is really no need to haggle over the respective merits or otherwise of different fire departments; the main point is, that they are, one and all, imbued with the same fighting spirit, and, one and all, are allied against the same common enemy. But in the following pages will be found some description of the New York Fire Department, and the writer is quite content to leave to the verdict of others, a decision as to whether or no the city is well protected against the fire fiend. Since, however, his is the honor of being its present Chief, he must be forgiven for stating at once that it is, as regards personnel and apparatus, unquestionably the largest in the world. And well it may be, for it is called upon to guard property values beyond conception, which from certain local peculiarities are, in a measure, heaped together in a limited area, and which offer to the fire-fighter problems, the solution of which are literally staggering in their immensity.

It is well nigh impossible to describe with pen and paper for the benefit of those who have not seen it, exactly what "New York" means. Most cities of renown convey certain vague impressions to those who have never visited them, and only know of them by repute or from what they have read. Thus "Rome" conjures visions of a bygone era of imperial greatness, manifested in wondrous churches, palaces and remains of historic interest. "Paris," speaks of a gay life, restaurants, pretty women, a continual effervescence of amusement with the serious side of life carefully hidden away in the background. "St. Petersburg" visualizes mentally, snow-capped domes surmounting fantastically constructed cathedrals, nihilists, an eternal carnival of disorder and, in the foreground, thousands upon thousands of gray-coated soldiery. As for London—without any intention of hurting the feelings of the British— fog, royal display and pomp, old and venerated buildings, art collections, military music and suffragettes.

Now, each and all of these figments of the imagination possess that grain of truth, which goes to show that years of descriptive writing have at least brought home to the public mind some regularly formed impression anent the places mentioned. Now, far otherwise is it with New York. As the foreigner, who has never been to America, what are his impressions of this great city? He would probably promptly reply, "giant buildings," and then pause. Perhaps, after a moment's consideration he might add with a half apologetic air, "and a very expensive place." Further than that he could not go, although viewed in certain aspects New York has the most strongly marked individuality of any city on the face of the globe. In the first place, it is as though some giant contractor had constructed a play-

ELECTRICALLY OPERATED HOOK AND LADDER TRUCK

thing for the gods, and had thrown vast piles of stone, mountains of brick, forests of wood and lakes of mortar on to the surface of one small island and had then fashioned therefrom buildings. There is no coördination, no attempt at architectural regularity; the wand of the magician waves and instanter a giant structure rises from nowhere, and in an inappreciable space of time a new skyscraper is silhouetted against the horizon. The sky-line alone is worth a trip across the ocean, so replete is it with fantastic wizardry and comparable only to the Organ Mountains as they loom out of the morning mist in the harbor of Rio de Janeiro.

But these aspects are rather for the brush of the artist; there is the other and practical aspect, which in its way is as enthralling. The population of New York is roughly five millions, and every day in the year one half of that number are on the move. They pour into a circumscribed area like water into a bottle, and as the number increases so, like the aforesaid water, they mount higher and higher as though they would overflow into the regions of the upper air. The office buildings lead the way in the race for height, and story by story climb five, six, seven and almost eight hundred feet in a search after floor space.

There are about one hundred and forty thousand separate manufacturing concerns in the city area and many of these are housed in buildings, sixteen and eighteen at a time, a perfect miscellany of diverse trades dealing in every conceivable article, inflammable, combustible or burnable. Hotels there are by the thousands, theatres literally by the hundreds, 900 moving picture shows, some 500 miles of docking and wharfage and innumerable ships and cargoes, the value of which it is impossible to assess, and, in any case, after a certain point the mind refuses to comprehend the meaning of recurring cyphers. But billions of dollars, millions of pounds sterling, may approximate to the fire risks which have to be covered by the New York Fire Department. Multiply this enormous aggregate of values by extreme climatic conditions which in themselves invite the attention of the flames, and the magnitude of

the task is enhanced. Then again multiply the result arrived at by the fact that business with a capital B is the
magnet which draws people to this centre; the hunt after
that elusive dollar which absorbs all the nerve power, the
intellect and the interest of the average individual, leaving
him no time for the consideration of the casual outside
occurrences of daily life, and tending to make of him a
machine rather than a man. That is where the carelessness of the unit may be expected to evidence itself, and
that, in itself, is one of the most comprehensive of fire
risks imaginable. Thus, it will be seen that the fire guard-
ians of New York have enough to do in their daily battle
against the fire fiend, and that personnel and equipment
must both be of the best obtainable if the enemy is to be
held effectually in check.

Firstly, as regards personnel, the force is recruited from
the State of New York, and, since apparatus without
skilled men to operate it is useless, the place of honour will
be given to the means employed to attract the best material
and how the service may be entered. The following form
the basic qualifications: "No person shall be appointed
to membership in the fire department or continue to hold
membership therein, who is not a citizen of the United
States or who has ever been convicted of felony; nor shall
any person be appointed who cannot read and write un-
derstandingly the English language or who shall not have
resided in the State one year immediately prior to his
appointment, or who is not over the age of twenty-one
and at the date of the filing of his application for civil
service examination was under the age of twenty-nine
years. Every member of the uniformed force shall reside
within the limits of the city of New York. Preliminary
to a permanent appointment as fireman, there shall be a
period of probation for such time as is fixed by the civil
service rules and no person shall receive a permanent appointment who has not served the required probationary
period, but the service during probation shall be deemed
to be service in the uniformed force if succeeded by a

permanent appointment and as such shall be included and counted in determining eligibility for advancement, promotion, retirement and pension as hereinafter provided." This is sufficiently exhaustive as showing the first main essentials to be covered by the applicant for fire service.

In addition to the foregoing, however, candidates must take a civil service examination, the physical and mental tests of which are as follows: As might be expected in such a calling as the fireman's, the physical test is in itself severe and searching. It may be said to be divided into two parts, the medical and the muscular development examinations. These are widely divergent, as the following shows.

For the former the candidate faces the doctors nude; prior to his entry into the examination room having taken his oath that he is whom he states he is, and that he has answered all questions which have been put to him truthfully. The applicant is then carefully examined for such defects as varicocele, hydrocele or any other kindred blemishes while, needless to say, any signs of venereal disease are met with peremptory rejection. Should he have any obstruction to free breathing, chronic catarrh or even offensive breath he may fail to pass. The teeth must be clean, well cared for, and at least twenty natural teeth must be present. In addition, any affections of the joints, sprains or stiffness of the arms or legs, hands or feet, ingrowing nails or hammer toes are especially looked for and promptly bar the applicant, since they would effectually prevent him from performing his duties in the manner demanded.

It seems scarcely necessary to add that rupture also receives particular attention, and any signs of incompletely healed laparotomy are noted against the applicant. Finally, the body must be well developed and nourished, and show careful attention to personal cleanliness.

Following this, the candidate next visits the first physical examiner, who tries his sight and takes his chest and other measurements. The minimum weight required is 140 lbs. on a height of 5 ft. 8 in. Then comes a stringent examination of the heart and incidentally, it is surprising, considering the age and physique of the applicants, how many have murmurs or some heart symptom, which in themselves probably not serious for most careers, are quite sufficient to prevent entry into the ranks of the Fire Department. This concludes what may be called the purely medical aspect of the case, with the addition of certain questions regarding past medical history, framed on a plan not unlike that drawn up by insurance companies.

The hour for the active physical test has now arrived. That it is severe may be gathered from the following particulars. Firstly, the strength of the upper arms is tested by the pull-up, or "chinning," as it is called. The candidate is required to hang in a suspended position from the rung of a horizontal ladder by his hands, and pull himself up till his chin is above the rung. A limit of fourteen times has been placed upon this operation, which would give a sufficient gruelling to most professed athletes. Next comes the "dip," or supporting the body by arms on horizontal bars and then lowering till the chin is level with the hands. This must not be accomplished more than six times.

A dynamometer is used to test the strength of the forearm, pectoral muscles, joints of the knees, back and legs. Considerable importance is attached to these readings. A sixty pound dumb-bell must also be lifted to the shoulder and thence above the shoulder with each hand in turn. Finally, the agility of the candidate is tested in a variety of ways, jumping being that most usually employed.

Now, it might be imagined that the only result of this extreme physical trial would be the elimination of all except giants and abnormalities, which, most assuredly, are not wanted. What is required is all round physical excellence, and to this end marks are given according to the stamina shown by the individual in the different tests and his familiarity with the same. But the outstanding point to be made is that, obviously apart from being in sound health, it is not by any means the strongest man who always makes the best rating. The youth who has

been gymnastically inclined will be familiar, of course, with such exercises as "chinning" and "dipping," and in that way will score; but it also shows that he has been in the habit of taking the best form of exercise and this in itself promises a deal. *Mens sana in corpore sano* might not inaptly be applied to the department, since the science of fire-fighting is a profession demanding brains as well as brawn.

As has already been stated, the mental examination is reasonable considering the training which is to follow, and consists of a thorough knowledge of the three R's—reading, writing and simple arithmetic. These examinations are, however, competitive and are rated on a basis of 50 to 50, i. e., half for physical and half for mental. It has been the practice, during the last few years, to appoint from these candidates in numerical order, although the Fire Commissioner has some discretionary powers under the law and may reject a certain percentage of the names certified for appointment. Unless a candidate has developed some serious defect between the promulgation of the list and the date of appointment, the plan of accepting men in numerical order is followed.

In order to detect whether such defect exists, the applicant is thoroughly examined by the medical officer of the Fire Department at the time he is about to be appointed. Should he pass all tests, he is appointed on probation for a period of three months and is immediately assigned to a company.

It is thought best to assign probationers to active duty at once, and in the heaviest and hardest districts of the city. A two years' service is required in these districts before a fireman is permitted to transfer to a lighter one. He is required to attend daily between certain hours at the college training school for probationers, there to undergo a thorough course of instruction in the use of scaling and other ladders on a building more than 100 feet in height. He is taught the use of every tool in the eight branches of the service—hook and ladder and engine companies, water towers, hose and chemical wagons, fire boats and so forth. Besides this he is instructed in the use and making of approved knots, such as the bowling knot, rolling hitch, the half hitch and others. Also he is instructed in the use of life saving apparatus and jumping nets. In addition he must learn and become proficient in sending and receiving all fire alarm signals.

In some countries it is thought best to instruct the men before sending them to companies for regular duty, and even officers as already stated in this volume attend fires as spectators. To the writer, this system does not appear to possess any advantage. In fact, practical experience has proved that it is better to throw the men at once into the thick of the fight. If they have any tendency to show the white feather this method speedily brings it out, but to the credit of the New York firemen, be it said, that in more than twenty years, during which time over five thousand men have been subjected to this ordeal, not more than ten have jibbed.

The reason why so few are found deficient in these tests is simple. They are aware of the conditions beforehand and this fact eliminates the cowardly and the timid. Once appointed to membership in the uniformed force, a young man can reach the highest position in the department, always, however, subject to competitive examinations. In these examinations record and seniority count fifty per cent., and mental qualifications the same. As regards the former rank takes precedence of service. A mark of 80 per cent. is given for the first six months in a grade, and one half of one per cent. for each six months thereafter up to eight years, when a full mark is given. This brings the rating to 96 per cent., the additional 4 points being reserved for merit marks (class A or 1), and a medal receiving three points (class B or 2), equalling 2 points, and class C or 3, equalling 1 point). These points are subject to change by the municipal civil service with the consent of the Mayor and State Civil Service Board.

The total strength of the uniformed force of the Department amounts to 4,995 men of all ranks, including, besides the Chief, 15 Deputy and 47 Battalion Chiefs, 11 medical officers, 298 captains, 413 lieutenants, 20 pilots, 496 engineers, 6 marine engineers, and 3,687 firemen of all grades. There are, in addition, 1 Chief of Construction and Repairs, 2 Roman Catholic and 2 Episcopalian Chaplains.

Upon being appointed to 4th grade firemen from probationers, men receive $1,000 per annum, being a year later advanced by law to the 3rd grade, with the same salary, and thereafter, by law to the 2nd grade, with $1,200 a year, and thence to the 1st grade, then receiving a salary of $1,400 a year. Thenceforth the progress of the individual is strictly dependent upon his success in competitive examinations. Thus, by taking practical tests, he may qualify as an engineer, with a salary of $1,600 a year.

A fire engine is taken to the water front and placed at the dock drafting water. Candidates are required to operate the engine in the presence of engineers representing the Civil Service, and are rated up to 50 per cent., according to the ability shown. A mark of 70 is required to pass this test. The mental test which follows is a written examination counting 50 per cent., a passing mark of 70 per cent. being required. When both tests are combined the candidate must have at least 80 per cent. to entitle him to be on the eligible list.

Or, again, he may follow the main stream of promotion, becoming first a lieutenant, with $2,100 a year, and after six months may rise to be a captain, with a salary of $2,500 a year. With six months' service to his credit in this grade, he is eligible as Chief of Battalion, in which position he takes charge of six or seven companies, and receives a salary of $3,300 per annum. Ranking with the latter are Doctors, Chaplains and the Chief of Construction and Repairs. Incidentally, it may be of interest to note that the Chaplains are provided with a departmental horse and buggy, for the purpose of attending fires, at which their services on occasion are greatly appreciated.

The next step of promotion is to that of Deputy Chief of Department, with whom rank the two veterinary officers and the Chief Medical officer. The Deputy Chief of Department in charge of the Boroughs of Brooklyn and Queens, owing to his greater responsibilities, receives a salary of $7,500, the Chief receiving $10,000.

Since the vacancies in each rank are limited, a list of candidates eligible for promotion is drawn up and remains in force from one to four years, when the list is closed. This is to enable the younger men to have their chance and to prevent their promotion being blocked indefinitely.

Without going too deeply into the subject of examinations, it may be not without interest to give two or three specimen questions taken from the examination papers for promotion to lieutenant, which will be sufficient evidence of the thoroughness and searching nature of the examination.   1. Assume that a fire has broken out on the 12th floor in the rear of a modern 20 story fireproof building, 100 feet deep, elevators 30 feet from the front. The room where the fire started is 30 by 50 feet, and is filled with a number of old desks, rugs, partitions and other furniture of a highly inflammable nature. Intense heat has developed before the firemen reach the scene, and the fire has worked its way into adjoining offices on the same floor.   (a) What are the special dangers to be apprehended from such a fire?   (b) What special precautions would you take to avoid loss of life?   2. In answering the following, candidates will show the methods used in arriving at the answer given. There is a large fire in a building 100 feet high from the curb to the cornice. The sidewalk is ten feet wide and the street from curb to curb is thirty feet wide. A pipe-holder is placed against curb on the opposite side from the fire. You are required to deliver an effective stream of water to the top floor. You are using 300 feet of three-inch hose with a 1½-inch nozzle. What is the approximate distance from the nozzle at the curb to the top story windows, allowing ten feet to a story, what pressure would you require on the nozzle, how many gal-

lons of water would the nozzle discharge per minute and what would be the required pressure on the engine or hydrant to maintain this discharge? 3. What is the duty of the commanding officers upon arriving at fires where the buildings have automatic sprinkler equipments? (b) After the first line of hose has been stretched in at a fire by an engine company and a second line of hose is required, what do the rules demand commanding officers to do?

It goes without saying that such technical knowledge cannot altogether be acquired by practice alone. There must be a sound theoretical training to amplify the latter, and to this end the Fire College was instituted in the year 1911. It may not inaptly be compared with the Staff College of an army, since all those joining it must be officers who wish to qualify along certain highly specialized lines. But the simplest method whereby an idea may be gained of its scope is to quote from the words of its charter. "To disseminate knowledge of fire-fighting, to establish and maintain the highest professional standards and to afford to men starting in the profession of fire-fighting the experience of men who have devoted their lives to the profession."

Included in the courses are: General fire-fighting, use of apparatus and tools, engines and boilers, high pressure system, marine fires, care of horses and hose, high tension electric currents, combustibles and explosives, gasoline and motors, fire alarm telegraph, auxiliary fire appliances, first aid to the injured, discipline and administration. Instruction is given by a detail of officers, but special lectures are delivered by distinguished professors from New York colleges, while the President is always the Chief of the Department.

As a company is the unit of action at a fire, and as collective work is of more value on these occasions even than in war, the Company School has been made a feature of the college work. From this it must not be gathered that the intelligence of the individual is not taken into account. Many times a man is thrown upon his own resources where he can neither see nor hear the signals of his superiors and care is taken in his training that his individuality and initiative shall not be repressed. The school is attended by companies, and complicated evolutions are performed. Each year a competitive drill is given, which is required to be accomplished within a specified time, and the men of the company most successful in time and form receive a college medal, while the Captain is given a special medal for the best drilled and most effective company.

Some idea of the evolutions in one of the recent tests can be gleaned from the following: There were twenty tests, which took place on and in a building 100 feet in height and of nine stories. (a) Stretch a three-inch line from high pressure hydrant. Connect to stand-pipe floor valve inside of the building; the outside connection is out of order. Winning time 41 seconds. (b) Raise and operate an aerial ladder. Winning time 50 seconds. (c) Hoist a line to the roof from the outside of the building. Make the line fast under the cornice and on the roof with approved knots. Winning time 1 minute 13 seconds. From these figures it can be gauged that the New York fireman is second to none in speed of operating apparatus.

There is in addition a large civil department, which is responsible for such auxiliary bureaux as the Fire Alarm Telegraph Bureau, which totals 133 men, including 30 telegraph operators, 14 battery men and 32 linesmen; the Fire Prevention Bureau which is charged with the making of inspections, the cleaning away of rubbish, etc., and includes within its scope the division of the Fire Marshal, who is responsible for inquiring into the causes of fires and for procuring evidence in cases of arson. In addition, there is a large Bureau of Repairs and Supplies, which totals no less than 251 employees, ranging from clerks and stenographers to painters and even a sailmaker.

The administrative head of this vast organization is the

Fire Commissioner, who is a nominee of, and appointed by, the mayor. In fact he may not inaptly be compared with the departmental secretaries of any national administration; that is to say, with a change of municipal government, *ipso facto,* he vacates his position.

Quarters are provided for all members of the uniformed force, at the stations to which they are attached, 4 hours daily being allowed for meals. In addition, every fireman is entitled to 14 consecutive days' vacation leave in the year, together with 24 hours each fifth day, while company commanders have it in their power to grant extra leave of 12 hours 4 times each month should the exigencies of the service permit.

The regulations regarding pensions are both comprehensive and generous. Any officer or member of the uniformed force who may, upon an examination by medical officers, be found to be disqualified, physically or mentally, from the performance of his duties, shall be retired from the service and shall receive an annual allowance as pension in case of total disqualification or as compensation for limited service in case of partial disability. In case of total and permanent disability caused or induced by the actual performance of the duties of his position, or which may occur after ten years' active and continuous service, the amount of annual pension allowed shall be one-half of the yearly compensation given as salary at the date of his retirement from the service, or such less sum in proportion to the number so retired as the condition of the fund will warrant. This fund is formed of certain revenues allocated for this purpose and any deficit is supplied by an appropriation from the city. In any case, widows and children of men killed in the execution of their duties are cared for, and members wishing to retire after a period of 20 years' service, receive a pension equal to one-half of their salary.

The City of New York is divided into five Boroughs, Manhattan, the Bronx, Richmond, Brooklyn and Queens, the two former being separated from the third by the bay, and from the fourth and fifth by the East river. For fire protection the Department consists of 298 companies equipped with 877 pieces of apparatus, including engines, hose wagons, hook and ladder trucks, fire boats, search light engines, water towers, etc. Motorization of all apparatus is proceeding rapidly and by the end of 1914 the horse will be practically superseded.

Of the operations of the Department, as a whole, some idea may be gleaned from the fact that between January 9th and March 1st, 1912, New York literally burned day and night. One period of 24 hours in February of that year witnessed a hundred calls. This tremendous activity on the part of the fire fiend was due to some extent to the extreme cold, the thermometer registering zero practically the entire time, but, judging from the revelations recently made, the firebug played no inconspicuous part in this state of affairs.

It is the custom of the New York Fire Department never to leave a fire station uncovered, and, in case of outbreaks of any magnitude, apparatus from distant stations takes up its position in the station vacated. This system will be understood by an examination of the accompanying card. All companies in the first horizontal line respond on the first alarm, i. e., when the street alarm is pulled by a citizen or a policeman. All companies and officers on the second line respond on the second alarm, which is transmitted in the following manner. Should the chief in charge at a fire find the outbreak to be beyond the control of the apparatus of the first assignment, he orders the transmission of a second or third alarm. This is always done by an officer or by an aide to the chief officer for, though the alarm box may be opened and pulled by a citizen, in order to send additional calls, an inner door must be opened to which none but officers or aides have access.

The mechanism within this door resembles the apparatus in use in a telegraph office, and the officer will send in his call by tapping 2-2-279; meaning a second alarm at station 279. A third, fourth or fifth alarm is sent in the

279 Broadway and Great Jones St.

ENGINE COMPANIES						H & L CO'S	DEP CHF	BATTALION CHIEF	B T	W T	TO SUPPLY FUEL	From Depot No.	COMPANIES TO CHANGE LOCATION — ENGINE COMPANIES	H & L CO'S
33	20	25	13			20	9	2	3	6		2		
72	55	18	30	24			3		5			W 3 2	5		
											W 1 11	5	16-17	26-14
9	27	7	29	15	3	11			4			W 8	5		21-3
14	31	5	17	28	11	5		2						
1'	32	6	19	4	21	8							54-1	10-6

same manner, merely changing 2-2 into 3-3 and so on.   It will be seen by the card that, on the third alarm, in this instance engine company 16 takes the places of engine company 17 in the latter's station, and engine company 26 fills up the gap left by engine company 14.   Similar changes occur on the fifth alarm, and hook and ladder company 21 "covers in" for hook and ladder company 3.

If the fire happened to be of such magnitude as to demand an even greater force than would respond on the fifth alarm, distant companies in the same Borough can be summoned by sending what is known as a simultaneous call, or what firemen term "the two nines."   The call is, to give an example, 9-9-279-3-3-582, which means that all companies due on the third alarm at signal box 582, would immediately respond with all apparatus to signal box 279. In lieu of this can be sent the borough call, which has exactly the same effect, but the signal is varied to denote the borough from which the apparatus is required, which, in the case of the Equitable fire, was Brooklyn.   In this instance it will be readily understood that there was no object in sending five miles north on Manhattan Island for companies, when the same force was just one mile away across the Brooklyn Bridge.   Should the borough call be used, it would be 7-7-279-3-3-394, indicating that the companies assigned on the third alarm at station 394 Borough of Brooklyn, would immediately respond to station 279 Borough of Manhattan.   6-6 denotes Manhattan, 7-7 denotes Brooklyn and 8-8 Richmond.

There are 3,176 street fire alarm boxes and, roughly, over 5,500 special and automatic signal boxes located in factories, public buildings, theatres, etc., and approximately 1,969 miles of underground conductors, while, by the end of 1914, the 2,530 miles of overhead cables will also be laid in the former way.   The fire alarm stations are situated as far distant from the scene of any possible outbreak as is practicable, those at present in the course of construction being located in the public parks of the various boroughs.   The Department is housed in 326 buildings, and every year sees an addition to their number. In 1912 there were 15,633 fires as against 13,868 in the previous year, the total fire loss for the former year being $9,069,580, as against $12,470,806 for the latter.

Occasionally, in the preceding pages, the reader may have thought that the writer was complaining of the position held by the firemen in the public esteem. Certainly it is true that the profession is of modern growth and, as such, has not behind it those centuries of glorious record which make citizens of every nation proud of their army and navy. But inasmuch as New York possesses the largest fire department in the world, and inasmuch as its uniformed force is called upon to combat a greater fire risk than elsewhere to be found, then most assuredly is this fact recognized by its inhabitants. It is no exaggeration to say that no force in the United States, federal, state or municipal, stands in more sympathetic relationship to the public. An English friend of the writer's, who knew the Department well, was asked what be. thought of it as a whole. He replied:

"Well, the greatest compliment I can pay is to put them on the same plane with the London police, whom I consider the most efficient, best mannered and withal the kindliest force it is possible to imagine."

It goes without saying that the writer is proud of his command, proud of their actions, which are engraved in the hearts of the people, and proud to think that Providence has permitted him to be their Chief.

CHAPTER XVIII

THE problems confronting fire departments in seaport towns in America are of a nature so widely divergent from those needing solution in Europe, that a few explanatory words are rendered necessary. As a general rule, tidal influences and depth of water play so important a part in the latter that it has been obligatory to construct docks which shall always be possessed of a certain depth of water. There is no need to labour the obvious point, that this has entailed harbour construction on a gigantic scale, involving in many cases the expenditure of millions of money. In this respect the Atlantic seaboard of the United States has been peculiarly fortunate, and it is possible, with rare exceptions, to berth the average steamer alongside a wharf projecting directly from the shores of the river, bay or estuary. This, of course, spells cheapness, celerity in dealing with cargo and a certain amount of convenience for passengers in transit. But, on the other hand, it has tended most distinctly to increase fire risks. In the designing and building of docks the greatest of care and forethought is naturally exercised in the safeguarding of buildings from fire, if for no other reason than the difficulty that must be experienced in successfully mastering outbreaks in congested areas, dependent upon their ingress and egress for the state of the tide.

Shallow draught fire floats have been constructed for this special purpose, but their capacity and radius of action are obviously limited, and hence any comparison between American practice and European methods is out of the question. The sheds, or wharves, common to America, form about the most dangerous structures of their kind in existence. Built on wooden piles, with wooden superstructures, they are comparable to nothing but horizontal flues, through which flames rush with a lightning-like rapidity, rendering abortive any efforts on the part of the fire department, unless the greatest promptitude is shown by all concerned, and demanding the use of fireboats with specially designed and extraordinarily powerful equipment. Fill these sheds with every sort of combustible material imaginable; hogsheads of resin, bales of cotton, crated furniture, barrels of pitch, stacks of drygoods, and such unconsidered trifles as a few boxes of celluloid toys and novelties, and can the mind of man conceive a collection of heterogeneous merchandise more calculated to provide the wherewithal for a conflagration and matter enough to assuage the thirsty pens of all the newspaper reporters in the town? Yet this represents an every day condition in an American port, and it is perforce necessary, not only to guard this property but to calculate the even more important risk, namely, should fire occur, the danger of its spreading to adjacent dwellings. Hence, even the inexperienced lay mind can easily grasp the vital significance of fire prevention under such circumstances. But incidentally, there is yet a further consideration demanding attention, the possibility of a fire occurring in a vessel moored alongside one of these piers. Fire risks on shipboard are appreciably greater in harbour than at sea. Discipline is relaxed and sailors and stevedores are human. After a hard morning's work a pipe of tobacco or a cigarette is a welcome solace to the most ascetic of individuals, and a carelessly thrown match, or the residue of a finished pipe is all that is necessary to start a blaze, which shall in one fell swoop destroy ship, cargo, wharf and men. In addition, though this may be scarcely credited, merchandise, particularly cotton, is often on fire before it is loaded, in which case it is absolutely a matter of luck where the outbreak occurs. Therefore, it behoves wharfmasters and

OIL FIRE.

captains to exercise the most stringent supervision over the goods they are handling. These are some of the complexities which face the master mind of the fire department, and be it remembered that no matter whose the initial responsibility, if a fire gets out of hand, criticism, and perhaps blame, will be apportioned liberally to the department whose services have been requisitioned to overcome the errors and carelessness of others.

To meet such contingencies the first essential is a flotilla of well equipped fire-boats, numerically sufficient for the demands of the harbour they are to defend. Much depends upon the architect chosen to design them, and he should be given a free hand and be untrammeled by petty restrictions, though needless to add, he should be a master of his craft, while too much emphasis cannot be placed upon the prohibition of untried innovations. These latter may result in serious loss of life and property. It is commonly held that fire-boats should be of the twin-screw type; this, as rendering them more handy for manœuvering in narrow crowded channels. But the twin-screw boat is more expensive to maintain and operate, and since economy is the watchword of municipalities, it has been found expedient to evolve a design, the turning circle of which with a single screw approximates to that obtained by two screws. Shorn of scientific formulæ, this consists in constructing the keel of the vessel from the midship section aft along a rising gradient, thus bringing the turning point well amidships, so that the boat can answer her helm almost as though she were on a pivot. In practice, this type of construction has proved eminently successful, eminently economical and in all respects satisfactory.

The writer has had little experience with the turbine engine as a method of propulsion. But at present, while excellent for driving pumps, certain difficulties over a satisfactory reversing gear render this system in its existing state of development useless for propelling fire-boats. Naturally, of greatest interest to fire fighters is the question of the pumps, their style and capacity; for as long as the driving engines are of the best compound type,—engines of this construction are more easily heated than those of triple expansion, and hence are of more general use in fire-boats—there is no particular specification of propelling machinery to be recommended. Many boats of old design are fitted with reciprocating pumps, and have done and are doing excellent work. But it is almost impossible to obtain boilers capable of operating the latter at their full capacity say of 10,000 gallons a minute. It resolves itself into a question of piston speed, and it has been found that under working conditions the steam supply estimated to obtain the same, as predetermined, rarely accomplishes its task. Now, centrifugal pumps can do the same work with half the steam, and hence, with care, this should enable a boat to manœuvre and at the same time to run her water battery to full capacity. The total volume of the streams per minute, in the opinion of the writer should amount to 7,500 gallons of sea water at a pressure of 185 pounds per square inch. As far as actual deck equipment is concerned, the main feature is the emplacement of two circular turrets, each operating nine separate streams of water and surmounted by a turret pipe, through which a powerful jet of from two to three inches can be thrown. The old style of running a water circuit around a boat represents considerable weight, causes confusion and adds to labour. The turret concentrates the work and is better adapted for the supervision of the officer in charge. Provision for a water tower can be arranged by constructing the mast on the military lattice girder system, surmounted with a fighting top platform, so designed that not only can two turret pipes be operated therefrom, but also a number of smaller jets, which may not inaptly be compared with the machine guns of the battleship. This system has been found most effective in fighting warehouse and pier fires. Boats should be kept under steam at all times in order that there may be no delay in starting. In New York, where there is a large fire flotilla, it has been found necessary to organize the same into a homogeneous unit, the better to insure its efficient coöperation. The Chief of this marine battalion responds to fires in a steam launch and commands on the water side. A code of signals has been established by means of the siren whistle, and in this way orders are transmitted from the Chief over a considerable distance, even though the smoke be heavy and the boats invisible. Above all, the personnel of these craft should be accustomed to the handling of boats, if not actually sailors. They must be alert, active and intelligent, while it is obvious for economical reasons that the engineers and their assistants should be highly qualified. Otherwise constant repairs will keep the boats out of commission and entail vexatious expense.

One of the most terrible disasters, which well exemplifies the perils of wharf fires, occurred on Saturday, June 30th, 1900, at the Jersey side piers of the North German Lloyd in New York harbour. At four in the afternoon on that eventful day, while hundreds of curious visitors were inspecting the four latest additions to the German line, a fire broke out amongst some merchandise on Pier 3, alongside of which the steamer Saale was moored. The origin of the outbreak is obscure, but it was probably caused by some unconsidered act of carelessness either on the part of an employee or of one of the sightseers. Be that as it may, in less than fifteen minutes the flames, fed by stores of cotton, turpentine and oil which were lying unprotected on the docks, swept with inconceivable speed from pier to pier, and before the immensity of the outbreak could be realized an area a quarter of a mile square had been devastated. A strong breeze was blowing from the southward at the time and to this the staggering rapidity of the conflagration was no doubt partially due.

Now, it must be clearly understood that, owing to a curious anomaly, the New York Fire Department, including its fire-boats, had at that time no jurisdiction in the State of New Jersey, and hence was unable to afford any assistance to the vessels in distress, until they were in the open stream and upon, so to speak, neutral waters. Then everything was done which science, skill and daring could suggest. In view of the disastrous turn of subsequent events, this fact should be borne in mind.

The first intimation of something wrong reached visitors in the Kaiser Wilhelm der Grosse, the then "ocean grayhound," through the appalling noise of hundreds of barrels filled with pitch exploding from the heat, like salvos of heavy artillery. A wild rush for the gangways followed, but the ship's officers with consummate coolness averted a panic by announcing that the vessel would proceed immediately into midstream, happily having a sufficient head of steam to accomplish this manoeuvre. The seamen hastily cast off, but so intense was the onrush of the fire that one man slackening a stern hawser found the wire already glowing from the heat. With her decks ablaze, her woodwork crackling, and clouds of steam roaring through her exhaust pipes, she presented a terrifying spectacle as she made her way slowly to safety. Tugs immediately went to her assistance, her guests were rapidly transferred and the fire was extinguished, but not before considerable damage had been done to her splendid and luxurious cabin appointments, which had been the talk of both sides of the Atlantic. Alas, not so fortunate were her sisters, who, not having steam up, were powerless to escape. To depict exactly what occurred upon these vessels at such a time of confusion and horror would in any case be almost impossible, and, in addition, of the few survivors, none could give a coherent narrative since, practically imprisoned upon the lower decks, they were able only to realize that death in some form was threatening them. The Saale, in the very heart of the flames, was cast loose and drifted slowly into the stream, a menace to shipping and a veritable funeral pyre to those on board. Hundreds of desperate creatures jumped overboard and were picked up by passing boats. But hundreds of others were less lucky and were roasted to death in the depths of that floating inferno. Little could be seen of their plight, but as fire-boats surrounded the smoking hull, faint cries from the lower ports attracted

DOCK FIRE, JERSEY CITY.

attention. Suddenly a naked arm shot out through the murk, and a voice, cracked with terror, screamed for help. Rescuers placed a hose line in the grasp of the quivering hand and as the water brought temporary relief, the crazed sufferer was understood to say that with him were forty odd men and women awaiting their doom. A desperate effort was made to haul him through the port, but his shoulders prevented his escape, and even as he was making one supreme attempt to dodge death, a wisp of flame shot wickedly out from behind him and branded him with its fiery tongue. With a shriek of demoniacal laughter, he surrendered himself to his agony and fell back to be seen no more. Another belch of smoke from the port and then a horrid silence. The little band of prisoners were beyond human aid, and had journeyed to that bourne from which no travelers return.

Slowly the Saale drifted down the Hudson, a moving emblem of the vanity of life and the evanescence of all things. Before it finally grounded off Ellis Island another incident replete with painful tragedy was to occur. A woman's voice was heard calling from one of the ports of the main deck cabins, and rescuers could plainly see and converse with its owner—a stewardess. Again the narrowness of the ports spelled death—a death so supremely horrible in its essentials that it scarcely bears narration. The fire was just eating its way through the paneled door of the cabin, and with the aid of a hose length from a tug the woman fought gamely for her life. Needless to say, the odds were all on one side. To escape was impossible, but none the less the unequal contest continued until, with hands blistered, eyes blinded, clothes burning, with a cry for mercy to her Creator, this brave soul passed to her reward. The case of the Main was equally desperate. Although a thousand feet away from the outbreak she caught fire almost instantaneously, and her decks were swept bare as though in the path of some giant tornado. There was no time even to cast off, and until the flames had eaten through the connecting hawsers she weltered in a whirlpool of fire. A few persons had jumped overboard at the first alarm and were seen clinging to her propellers; then one by one they were overcome and dropped off in the muddy eddy which lapped the dock wall. Thus she lay for some hours, and it is indeed surprising to relate that even sixteen persons out of her complement of one hundred and fifty managed to escape. And the story of their escape is indeed miraculous. They were all coal-passers or engineers, engaged in their professional duties about the engine room. Upon the alarm being given they had found all means of exit to the deck cut off by the flames, and had consequently retreated to one of the coal bunkers, the door of which they closed. For eight hours they remained there uncertain of their fate, ignorant of what was happening and in a temperature which made it painful to breathe. For some considerable time the electric lights burned steadily, a sardonic reminder that a supernatural stoker had taken charge of their duties and was generating the steam necessary to keep the dynamo running. Then came a flicker, the lights shone with an uncanny brilliancy, and then there was darkness. The silence was still broken by the monotonous hum of the ammonia pumps, connected with the refrigerator plant. It seemed as though their speed were being increased by some ghostly mechanic, for the hum developed into a mighty roar, culminating in an explosion, and then there was silence.

It was at eleven thirty that night that the poor, helpless Main was grappled by a fire-boat, the crew of which, hearing voices, located these prisoners and succeeded in hauling them one by one through a coal port. Their condition was desperate, but ultimately they all recovered, with the exception of one man who had been partially blinded by steam and died in hospital.

The plight of the fourth liner, the Bremen, was not quite so critical, as the fire did not succeed in getting a good hold below decks. She was, however, crammed with visitors, which would account for the fact that seventy-four persons perished aboard her. Like the others she drifted away from the burning docks, and it was some considerable time before tugs and fire-boats had succeeded in getting her under control. Meanwhile, she was acting as a veritable torch to all shipping and wharves with which she came in contact. Carried by the current toward the New York shore, she imperilled all the docks from Thirty-third Street to the Battery. In fact, so serious was the menace that alarms were sent in to city fire stations, and men and apparatus stood by, ready for all eventualities. One lighter passed her, caught fire, and drifted alongside the Baltimore and Ohio wharf, which promptly in its turn took fire. Fortunately this outbreak was quickly sup-pressed, but the same thing occurred at several points, a sufficient indication of the peril which was threatening the whole river front. However, the danger was averted and the Bremen secured and later beached in shallow water.

On the Jersey side of the river, the desperate work of the fire-fighters had had its effect, and the Scandinavian-American Line docks, which adjoined the North German Lloyd, escaped with the inevitable injury caused by burning embers starting subsidiary fires. At one moment, it was seriously feared that they, the Hamburg-American Line, the Holland-America Line and the Wilson Line sheds would all become involved, together with the vessels moored alongside, which would have constituted one of the great-est disasters in the history of maritime conflagrations. Happily, however, such a catastrophe was avoided, and in spite of the enormous damage the fire was practically under control within six hours of its inception. But to the day of their death, those who saw the Hudson in that summer twilight, will never forget its fantastic appearance. The four great liners vomiting flames and smoke, and sur-rounded with puffing tugs and busy fire-boats, while a couple of dozen smaller craft floated hither and thither on the most congested waterway of the world, aflame from stem to stern, and reminiscent of nothing so much as an

Armada of old time fire-ships intent upon destruction. Doubtless, human forethought, energy and determination in no small degree vanquished this enemy, but Providence must have been watching over New York that day.

The actual extent of the pecuniary loss entailed by this conflagration has been assessed at $6,000,000—a mere bagatelle in comparison with the four hundred lives which were sacrificed. There have been, of course, bigger disasters of the same nature financially, such as that of Hamburg where it is estimated that fire destroyed $45,000,000 worth of property, but none has approximated to this in its sheer horror, and in bringing home to the lay mind just what may occur as the result of a small outbreak upon a wharf. It is almost as though human nature required the sacrifice of life, grief-stricken homes and the poignant realization of the grimness of death, in order to bestir itself towards the adoption of fire prevention in its most simple forms.

Even the uninitiated will realize readily that the methods of coping with fires on board ships must differ radically from the systems commonly in vogue on land. In the first place the construction of a ship is such that successfully to deal with an outbreak bespeaks a rough general knowledge of naval architecture, without which the most intelligent officer must be hopelessly nonplussed. But under any circumstances, it is the business of the Fire Chief upon arrival alongside the vessel to consult with the Captain or whoever may be in charge, with a view to ascertaining, if possible, the location of the fire and the nature of the cargo, upon which latter much depends. In addition, on all large steamers a plan of the vessel is placed in the chart house and this will show in diagrammatic form, the various holds, with distances from one bulkhead to another, the ventilator shafts, and other details of the greatest value in deciding upon the plan of attack. The location of the outbreak having been ascertained, which should only occupy a few minutes, all hose lines should be stretched and in readiness before removing the hatches. Also, if steam is being already used to hold the fire, a common method on board ships, it should be shut off as soon as the preparations outlined have been made, as it seldom happens that men can enter either the holds or " 'tween-decks" of vessels where resort has been made to this plan, without allowing time for the atmosphere to cool; a matter of moments perhaps, but in such cases it is the seconds saved which count. Everything will now depend upon the location of the blaze. If it is in the lower holds, the best thing to do is to remove a hatch in the lower deck, drop the hose line through the opening and simply flood the compartment. Should the fire, however, be " 'tween-decks," a different means of attack may be successfully employed, providing there be port lights in the ship's side. These should be stove in about twelve feet apart, a fire-boat should be run alongside and should bring into play her lines armed with distributing nozzles, which latter should be forced through the ports, water being pumped in at a pressure sufficient to give about fifty pounds to the sprays. If it is possible to reach the ports on the other side of the ship, similar tactics should be adopted, and in most cases the fire will quickly be under control. Then a ladder should be placed down the hatch. As a rule there is a built-in ladder in every hatch, but failing this a regulation fire ladder should be used, providing solid foundation can be discovered for it. Further, all men employing the same should have a line around them, in order that they may be hauled to safety in the event of any accident. Ventilators leading from the deck on fire should be utilized for dropping down hose with distributing nozzles into the affected area, which will render valuable assistance in cooling down the compartment. Finally amongst preliminaries, if the pipes used for sub-cellar work are long enough to operate, they should be utilized down the hatchway. An important point to be remembered now is, that all cargo ports must be closed. This is rendered absolutely necessary, since the water being pumped into the vessel is bound to give her some list, and if the ports are left unclosed there is the strong possibility that she may heel over and consequently fill and sink. But this by no means concludes the long list of precautions to be taken or knotty problems to be solved. Since such great volumes of water are being steadily and persistently pumped into the ship, it stands to reason that her draught will rapidly increase, and if she takes the ground, she will instantly list heavily and probably endanger the lives of everyone assisting on board, let alone rendering the actual fire-fighting ten times more difficult. To keep her on an even keel is the primal necessity of the situation and this demands as much scientific diagnosis of the needs of the moment as ever medical man was called upon to expend over an unknown patient suffering from an obscure complaint. To those who know them, ships are almost human in their idiosyncrasies, and the slightest mistake in their treatment may spell irretrievable disaster.

First and foremost, if the vessel seems likely to take the ground, by hook or crook, get her off into deeper water, and should she be light, fill her ballast tanks. An expedient at times resorted to, but by no means to be recommended under ordinary conditions is to flood the life-boats on the weather side, thus so to speak levering her back into position. But this is obviously dangerous in the extreme and should never be resorted to, unless those superintending the operation are experts and understand shipcraft from A to Z. Again, should the fire be gaining ground and it seem as though it were getting out of control, it is impossible to avoid heroic methods, and she must be towed to shoal water and beached, care being taken that her decks will no more than lie awash. Admittedly this is a last expedient, but it will save her from total destruction, providing she is sunk in shallow water, which will, of course, make it possible to pump her out and float her again. It might be imagined that such total immersion would subdue any fire known to man, yet the fact remains that cotton is so obstinate in its resistance that the writer has seen bales, which have been a whole week under water, at a depth of forty feet, that, upon being examined shortly after coming to the surface, were not only smouldering inside, but, upon being prodded, burst into flame. This gives some idea of the stubbornness to be encountered in dealing with some cargoes, and it is small exaggeration to hazard the statement that raw cotton requires as much attention as guncotton, from the skipper's point of view, i. e., the safety of his crew and himself.

A vessel reaching port already on fire, and which has signaled for assistance, offers again a rather different aspect of affairs with which to cope. In this instance, steam should be kept playing on the affected area and the hatches battened down, until all passengers have been taken off. Anything likely to cause a panic would be fatal and quite unnecessary. Under the conditions named, there would be no danger of an immediate and fatal spread of the outbreak, such as in the case of a building might cause loss of life within a few minutes. As mentioned elsewhere in this volume, it cannot be emphasized too strongly that, in the opinion of the writer, steam alone will rarely extinguish a fire. In itself, it has already absorbed a great quantity of heat and its transformation from a liquid state into vapor has been due to just such elemental activity as it is now called upon to subdue. Hence how can it be expected to exercise a cooling effect, which, after all, is what is needed, when itself over boiling point. All that can be expected is some temporary check, consequent upon moisture, but as a permanent and real stay to flames it is comparatively useless. It seems almost needless to say that in bringing fire-boats alongside steamers unmanageable on account of fire or whose steering gear is in danger owing to its becoming affected by great heat, the former should take up a position on the quarter from which it is possible to control and steer the latter. In addition, in all open waters, care should be taken so to handle the burning vessel that the flames may be prevented from sweeping the decks, as undoubtedly would occur were she forced head on into a strong breeze, the fire being forward or vice versa. In either case she should be kept before the wind, thus minimizing the area open to attack and at least giving the operators some deck room upon which to organize their defense. Though strictly, perhaps, not within the scope of this chapter, the writer is strongly of opinion that the time has arrived when all ships should be compelled to carry some simple and effective form of automatic fire preventive apparatus. The sprinkler system would appear to offer many advantages and to be easy of installation in vessels of new construction. This might be controlled either from the chart house, where exists at present the "smoke pipe" designed to warn the officers of a fire in any hold, when upon being definitely located the system might be brought into operation, the flow being controlled by the ship's pumps, or alternatively the installation might be arranged on lines broadly similar to those in use at theatres. This would demand a fusible plug, which at a certain temperature would melt, allowing a heavy and constant stream of water over a certain defined area, the pressure, of course, being constantly maintained by the ship's pumps as in the other case. No doubt, expense would be urged as a deterrent to the introduction of any such appliances, but it does seem passing strange that when precautions without number are now being taken to save the careless from the comparatively rare peril of the iceberg, so little attention is given to the ever-present menace of that most ghastly enemy at sea, the flames. Or can it be, as has been suggested, that what is wanting is the lurid lesson of a great fire in mid-ocean. In conclusion, a few words may not be amiss anent the position the fire-boat may conceivably play in any municipality boasting of a water way, without which the necessity for such a costly accessory could not exist. It is commonly presumed that a fire-boat, as such, must confine its attentions to its own element, and can in no wise be regarded as amphibious. This is an error, which has been practically demonstrated by the writer. Properly handled, the fire-boat becomes a most powerful and useful auxiliary to land apparatus.

During the San Francisco conflagration, it is reported

James Duane IN ACTION.

that from a Government Revenue cutter a line of hose was run for half a mile and that its coöperation even then was valuable. The words, "even then" are inserted, since with the limited pressure available from the pumps of such a vessel and with no natural aids such as gravity, the nozzle power of such a stream could not be seriously considered as of particular importance, unless water was altogether lacking, as in the case mentioned. But from tests made in New York it was conclusively proved recently that it was feasible and caused no undue strain on apparatus to discharge a jet through a one-and-an-eighth-inch nozzle at about two-thirds of a mile from the fire-boat acting as a pumping station, the nozzle pressure approximating fifty pounds to the square inch. There were two relay engines in the shape of two ordinary steam fire pumps, and when the pressure on the fire-boat registered 280 pounds, the further engine maintained a nozzle pressure of fifty-nine pounds, giving 291 gallons a minute, not a great stream, but considering the conditions of the experiment sufficient to show the possibilities attendant upon the introduction of the fire-boat as a land auxiliary. The distance for effective relay of water can be proportionately increased by multiplying initial fire-boat lines and siamesing them. One New York fire-boat can furnish twelve three and a half inch lines, sufficient to supply twenty-four engines under conditions similar to the test. That in itself is sufficient for the handling of a large fire. Thus, it will be seen that, in great emergencies, here is an auxiliary to the fire force on land which is at least impervious to the breaking of mains, climatic or seismic interruptions, and hence not lightly to be neglected.

CHAPTER XIX

FIRE STRATEGY IN THE HOMES OF THE PEOPLE

To prevent a fire is one thing, to fight it is another, and since this volume deals with all aspects of the subject some consideration must be given to "fire-fighting" in its active sense, that is, how to deal with outbreaks when an alarm is turned in and the best method of using apparatus. This may sound to the lay reader stale and unprofitable reading, but if he will take his courage in both hands and dip into the subject in ever so small a degree, he will find much to give him thought, much that will be of value to him and a certain number of useful hints which should be of assistance to him should he ever be unfortunate enough as to be involved in a fire. The fire risk becomes more apparent as soon as it is realized that it is a menace to the "Home," and that the flames are no respecters of persons and are as likely to visit the mansion or the tenement as the factory or the warehouse. It is hard to realize when reading newspaper reports of some great conflagration, quite what it all means; it seems so far away, so remote from the happenings of daily routine, that it is perused with passing interest and forgotten. Then comes the day, when suddenly the menace appears in all its lurid horror, and behold the occasion when an ounce of knowledge regarding fire and its usual course of progress may be the means of preventing the advance of the enemy and of saving human life. Of course, prevention is better than cure, but even as the hypochondriac occasionally falls a victim to the ills the flesh is heir to, and has to invoke scientific aid in order to regain his health, so is it with fire. With all the precautions in the world, it is impossible to guarantee that an outbreak will never occur, but it is within the ken of man, what then to do, and though the professional will be needed to fight with the sufferer as the doctor does with his patient, the individual who is prepared for the onset and knows just what course the attack will probably take is doubly armed against the foe. It is for this reason that the writer is hopeful that the general reader will persevere and glance through the following pages. Certainly if a person of discernment, the previous chapters will have shown him how life has often been unnecessarily sacrificed at fires, when well directed action would have saved the situation.

In all countries responses to fire calls, particularly those turned in from street boxes, are usually very prompt. As a rule about twenty seconds is consumed in hitching up and getting the apparatus out of quarters. Horses are trained to come to their places at the pole on a gallop and all harness is hung with an open collar, which locks with a snap. Commonly, three engines and two hook and ladder companies are designated to respond on the first alarm, and, in making the assignments, the company, nearest the spot from which the call was sent in, is expected to be the first to reach its destination. In forwarding reports of operations at fires the officers must state the order of their arrival and also which company gets to work quickest.

It will be easily deduced that this creates a spirit of the keenest rivalry, and to be beaten in a dash to an outbreak at which a company is first due is considered a humiliation. If it happens twice in succession the commander of the defeated company is asked for an explanation, allowances being made for gradients and traffic, and should it appear that the company is lacking in energy or vim caused in any way by indifference on the part of officers or men, a reorganization is sure to follow. This, then, is the primal step towards the formation of a good company, which is the unit of organization. Horses must be carefully trained, for a few seconds lost in getting away from quarters cannot be made up en route, particularly when the other companies have had a better start. The driver is as important as his animals and should be a man trained to the hour and one who knows every inch of the streets, and every fire hydrant in the district to which his company may be called. Many an occasion can be recalled by the writer when, on the way to a fire, he reviewed the whole district surrounding the box whence the alarm was received, and had the hydrants clearly in mind as the engine approached its destination; often pulling up in front of one completely shut off from view by obstructions, secure in his knowledge of its existence. The engineer should be equally well informed and should know the size of the main on which the hydrant is situated and how much water can be drawn therefrom. This prevents the mistake of having too many engines located on small mains with low pressure. When approaching the fire the engineer should jump from his engine, run ahead and have the cap off the hydrant as the apparatus pulls up. With the change from horse to motor as a means of propulsion, the department has been shorn of some of the spectacular features which in the old days added to the picturesqueness of the proceedings, but the dangers of injury in responding to a call through crowded thoroughfares are greatly increased. Stretching in hose lines at fires has been the subject of many orders and lectures, but when celerity is paramount, hose cannot be measured by the foot as can be done in a class room. However, when the location of the fire has been ascertained, the commanding officer should have a fairly good idea of the number of lengths required, and it is slovenly and shows but poor judgment to see hose coiled in the streets in front of the site of the outbreak. Controlling nozzles should be used for inside work in order that the stream may be shut off from this point, and at fires in tenements and private dwellings the greatest care should be exercised in the use of water, as the floors are not "filled." In such cases water runs through the floor boards, destroying the ceilings beneath and the furnishings

RESPONDING TO AN ALARM.

of the lower floors. The best way to illustrate these rules is to assume an outbreak in an old style tenement, where the blaze has started in a store on the ground floor. In addition to the main entrance and show window, the store has a door opening from the rear into a hallway where are situated stairs which lead to the apartments above. In this type of building there are usually six floors, including the store, with from three to four families on each. The stairs are of wood, and all partitions are stud covered with wooden lath and plaster. •The ingenuity of man could scarcely devise a better fire trap than that which is afforded by this style of construction. To prevent the fire from reaching the stair well, and by this means ascending to the top only to "mushroom" on the highest floor, is the first thing to be done. A line of hose should be taken into the hall to cover the rear door and transom to stop the fire spreading up the stairs, whereby that means of escape from the upper stories would be cut off. The second line should attack the fire from the front through the main door, while ladders should immediately be raised to the front fire escapes and the floors over the store should be examined to make sure that the blaze has not mounted through pipe recesses or other vertical openings. Should the alarm have been sent in with promptitude a fire of this nature can, and in all probability will, be confined to the store where it originated.

But if there has been delay and the fire has burned through the doors communicating with the hallway and stairs, and by means of these latter has swept through a crowded tenement, a very different problem awaits the officer in command upon his arrival. As in the former instance, the first line should operate in the hall to drive the fire back through the door leading to the store, and when that has been accomplished the nozzle should be directed up the stair-well, which is usually about twelve inches wide. A stream sent up this opening will deaden the fire down on the upper landings which, at present, it is impossible to reach in any other way. The second line

should be opened on the fire in the store from the front
and the flames subdued to prevent them ascending on the
outside of the building and enveloping people who, by
this time, would be crowding on the fire escapes. A few
seconds should suffice to quell this portion of the outbreak, when the line should be taken to the hall in order to
cover the first company which, by now, should be ascending the stairs. In the meantime, hook and ladder companies
would have placed ladders on the street side of the building
and would have begun the work of rescue from fire escapes
and windows. In cases of this nature events move swiftly,
and in far less time than it takes to write a line would be
rushed to the top floor by front or rear fire escapes, generally by the latter as the landings are, in most cases, nearer
the back of the structure. This line is of the greatest importance, and if the position is reached quickly the fire
should be prevented from spreading on the top floors, One
man should be sent by way of adjoining buildings to open
the bulkhead door, which is found in all places of this type
and which leads from the highest story to the roof, in
order to assist in accomplishing this purpose. All possible
effort should be made to force the advance line up the
stairs, and here is where men must take their medicine,
if need be. Life depends on their grit, determination
and endurance, for work of this kind often entails severe
trial. By this time the second line in the hall should have
killed the fire in the store and should immediately follow
in the rear of the other line on the stairway, in order to
prevent the flames coming out behind them and cutting
them off or rendering their efforts abortive. It is essential
in fire-fighting, as it is in battle, that companies or columns
be supported, or in firemen's parlance, "covered." To detach a column from the main army and direct it to attack a
well intrenched enemy without proper support would seem
to the writer to be a tactical error; and similarly, to order
an unaided company into a position where its retreat is cut
off by fire would be a very dangerous proceeding. In war
the column would, in all probability, fall into the hands of

the enemy and become prisoners, but fire takes no prisoners, and the battle cry is, "Death or Victory." Therefore, it will be understood that companies must support each other, the most advanced being always assisted and covered by the other coming up the stairs or on the floors below. There is one more essential point to be guarded, namely the light shaft which is usually between two of these buildings, and the windows opening into this space should be protected in order that the fire should not extend upwards on the outside or cross the shaft and ignite the adjoining buildings. This disposition of forces should successfully deal with a fire of such a character without loss of life, for during the operations described the two hook and ladder companies would have been engaged in conveying all the residents to a place of safety. Every action or order cannot be fore-told, nor would the writer attempt to say in what respect a commanding officer might find it necessary to deploy his forces, but he should use all possible efforts to outflank the fire. He should never neglect, when the flames have control of the stairway, to take a line to the top floor by way of the fire escape. In outbreaks of this character there is always danger of loss of life and the risk is great-est on the top floors, but with a line following the fire and a stream above it, the enemy is placed between two bat-teries and his doom is sealed. The opening of the bulkhead door in the roof allows the heat and flame to escape, gives the men coming up the stairs a living chance and permits the other company to effect an entrance from above in order to finish the work already well in hand. In dealing with an old style tenement it is important to look carefully around all pipe vents, and also the partitions near folding doors, as every stud partition is a vertical flue and there are few fire-stops. The lath should be stripped until it is satisfac-torily proven that there are no nests of fire concealed behind the plaster. Above all things, a search should be made as soon as possible in every room for persons who might have been overcome by smoke, and the floors under the beds should be carefully inspected for, as already stated, children will invariably take refuge in these places, and if not found and carried at once to the open air will die of suffocation. This is a brief outline of a fire, presupposed to have started in a store, but it may have originated in the cellar, when as the building has an inside stairway, this latter should immediately be covered to prevent the flames from mounting. Should it succeed, the outbreak is similar to the one described. Therefore, the first object as stated above, is to confine the blaze and then the cellar should be entered by the front or rear doors. When the outbreak has been located, distributing nozzles should be used through the floor over the section which shows the greatest heat. Rapid action is the keynote of success, and as fire spreads like lightning through these old buildings, a few seconds delay may mean loss of life. By quick stretching in and with good engines to give water to the lines the fire will be suppressed, and it is just these qualities of prompt action and swift decision which have saved thousands of persons in New York City. Each year ten thousand fires occur in these homes of the working population, making 64 per cent of the total number of outbreaks recorded. Much more might be written on old style tenement house fires, though should such an occurrence originate on any floor and spread to the stair hall the effect would be the same and should be combatted in a like fashion.

It is not the intention of the writer to lay down specific rules or to attempt to tell the Chief in command at a fire that this or that should be done, since local conditions alter cases. For the same reason, no rule is given for life saving beyond the raising of extension and other ladders on the front, and, if it be possible, on the rear of the tenement, in order to remove the occupants at different points. Extraordinary cases of peril, when all human ingenuity seems exhausted, often result in daring feats of rescue accomplished in a way that no one could foresee. In new style tenements the stair halls are enclosed in brick walls, the ground floor is fireproof and the hall doors leading to the different apartments are kalomine. There is usually a long private hall running the full length of individual flats which, should a fire break out, would soon be heavily charged with heat and smoke. Should the fire originate at the end farthest from the door through which the firemen enter, great hardship is entailed in crawling the length of this narrow passage dragging a line of hose. Water should not be started until the blaze is located, so that the stream may be played directly on the flames instead of doing unnecessary damage by opening the nozzle when no fire is visible. In order to relieve the men with the line, hook and ladder companies should be active in ventilating the apartment, and if there be a fire escape on the front or rear, the windows should be opened. In circumstances where the fire has gained considerable headway a second line should be taken up one of the escapes. Should a heavy flame belch forth from windows on the lower floors, there is always danger of the heat breaking the glass in those overhead, and in this way communicating the fire from floor to floor. This should be met with heavy streams from the outside to drive the fire back, using a water tower, mast, deck or turret pipe, or if none of these be available a street pipe may be employed, though these should be shut off when the flames have ceased to spread on the exterior. To locate and confine a fire to the smallest possible area is one of the first great principles of fire-fighting. As it is the tendency of flames to burn upwards, care must be exercised and attention given to the floors above the outbreak, though it should be understood that the floors below should not be neglected.

Fire-brands often fall through an elevator shaft, or other vertical openings, and start a blaze in the story below, so that attention must be given to all parts of the building involved and even to adjoining houses. Before leaving the scene of an outbreak the greatest care should be taken to see that all fire is really extinguished, since sometimes there is a tendency to overlook odd corners, resulting in a recall to the scene later in the day, an incident all fire-fighters regard as in some degree a reflection on their skill and often causing embarrassing questions. A new style of apartment house is being erected in New York City and elsewhere, divided into so-called "Duplex" flats. Each of these has one main entrance from the stair and elevator hall, while its upper floor is reached by an inner staircase with no means of direct exit to the main hall outside. In other words, it resembles a two-story house within another house. There are no outside fire escapes, no means of reaching the upper floor except by way of the inside stairs, and, should a fire occur, the heat and smoke would immediately ascend and all escape from the upper floor would be cut off. Firemen would be obliged to enter the apartment by the one door leading to the elevator and would find that the flames had already extended to the floor above. Two lines would be stretched either up the stairs or, if it be a building more than eighty-five feet high and the fire is above the fourth floor, from the stand-pipes, the engine being connected to the siamese at street level. The first line should be opened on the fire to clear the way to the inner stairs, the second line should also be brought into play to kill the fire on the lower floor and to cover the first line which should try to ascend. This may be almost impossible, as the heat and smoke will be confined above, unless the outbreak is of sufficient magnitude to melt the windows or blow them out, and if this occurs the contents of that apartment will be destroyed. To ventilate the upper floor is the problem before the officer in command and the only method of so doing is to use the hook of a scaling ladder or a telescope hook which has recently been designed for just such cases. When this has been done it will be possible to get the lines up the stairs.

Should the fire occur on the upper story it would back down and charge the lower floor with smoke, with a similar result. This type of building, while admittedly convenient for tenants, is most dangerous in case of fire, and people may be trapped twelve stories above the street. Personal risk can be obviated to a great extent by connecting each floor by means of a balcony fire escape on the inner court.

In cases where it is impossible to attack the upper floor by way of the stair, the terra cotta partition dividing the apartment from the hall should be broken through on a level with the upper story. This is not a very difficult matter and a hole large enough to admit a man would soon be made, and a charged line should be introduced. After the room is cooled it may then be possible to effect an entrance, when the line should be directed towards the stairs in order that communication with the floor below may be reëstablished, and a thorough search for possible occupants should be made.

It is remarkable how much more latitude is allowed in designing private dwellings than is given architects in plans for tenements, apartment houses and hotels, and many disastrous fires, often attended with loss of life, occur in such structures. A common defect is the large open stairway, and elevators have recently been introduced reaching usually to the top floor and capped with a four or six inch stone slab, which, from the standpoint of fire control is an extremely dangerous form of construction, as, should an outbreak occur, the flames will quickly ascend the shaft and having no outlet through the roof, will inevitably mushroom on the upper floors, thereby entailing possible loss of life. Such a thing as fire in any part of a house except in the furnace is not thought probable or even possible, so when the unexpected happens great masses of draperies, carpets and heavily upholstered furniture soon turn the building into a roaring volcano of flame. Many dwellings of the wealthy stand in an enclosure, often surrounded by high iron railings, making it particularly difficult to get any kind of ladder near enough to be of service. This class of structure has no outside fire escapes which increases the peril of people trapped on the upper floors. In such a case a person should never attempt to descend the stairs, but should, if possible, reach a front window, or any that can be found smoke free and call for help. Every minute of waiting may seem an hour, but some one in the neighbourhood will assuredly turn in an alarm. Above all, remember that jumping is the last resort of the desperate and, if driven to the final extremity, sheets, curtains or blankets will provide some sort of assistance towards reaching the ground.

On his arrival, the commander of the fire forces should at once stretch a line, taking it in through the main entrance and driving the fire away from the stairs. If, by this time, the flames have forestalled such an attempt, all speed should be used to follow in its path and should there be a back stairway this should be covered in the same manner. It takes but a few seconds to make a survey and to see in which direction the fire is spreading, when it should be outflanked with all rapidity and its progress stopped. The building should be at once "covered" with ladders, for while no persons may be visible some may appear at any moment. An extension ladder should immediately be raised, if the fire has control of the stairway, in order that a line may be taken through an upper floor window to cut off the advance of the flames. Men should be sent to the upper floors to ventilate the rooms and to search for persons who may have been overcome. The size of the house and the extent of the fire should be the guide to a Chief on his arrival in summoning additional aid, and no time should be lost in sending in further alarms where conditions are dangerous. It is earnestly suggested that owners should equip their homes with a few fire extinguishers to be used pending the arrival of the firemen, and it may be added that draperies are a prolific source of danger. The first floor at least should be fireproof, and in case of an outbreak in the living rooms, curtains should be torn down and the fire trampled out or beaten with rugs, using any means to hold it in check until the arrival of the trained fire-fighters.

Telephones as means of sending in alarms are untrustworthy, as under such circumstances excitement is liable to cause the message to be misunderstood, and in the case of town houses there is nearly always an alarm box in the vicinity.

The most dangerous fires in private dwellings usually occur at night after the occupants have retired, and as the bedrooms are generally on the upper floors the premises should be thoroughly searched the last thing at night by the owner or some responsible person, and if electricity is used as a means of lighting the wiring should be frequently examined. Both owners and architects are loth to extend an elevator shaft above the house, because of the inartistic effect of having an unsightly object protrude through the roof, but the shaft could be connected by a fireproof flue to one in the chimney which would carry smoke and fire over the roofs and ventilate the floors, thus giving the occupants a chance of escape. There also appears no reason why the sprinkler system should not be installed in houses which possess elevators, in order that in the event of a fire, sprinkler heads might operate in the shaft and prevent the upward trend of the blaze. In all the foregoing classes of buildings fire is always necessarily a great menace to their tenants and deaths may result from the smallest of outbreaks in a crowded tenement or in a great mansion. It seldom happens, however, that the life lost is that of a fireman, with the possible exception that he may be suffocated by escaping gas in cellars. Here lies one of the greatest dangers to the force, and gas should promptly be shut off from the building involved, especially when the fire has begun in, or has spread to, the basement. The above instructions are not difficult to comprehend or to carry out, that is to say as applied to the layman, all that is demanded is a minimum of coolness and the avoidance of panic amongst strangers who may be on the scene and are unaccustomed to the realization of the dangers of unconsidered action. Of course, in large country houses remote from any professional fire-fighting force, a private brigade should always be formed from workmen on the estate. It is not difficult to instruct them in the efficient use of simple apparatus, while prompt and timely action would in many cases save fine old mansions of historic interest, which alas! are annually lost by fire. Granted their construction is faulty and that the architects of three centuries ago neglected to consider fire risks, still even so, it is an error to give up the battle as lost before firing a shot.

CHAPTER XX

"QUICK BURNERS"

•

IN Europe no less than 350 years have been spent in building some churches, while in America 350 churches are built in one year.

There, in succinct form, lies the cardinal difference between European and American construction, and it is this latter which possesses a profound significance for the fire-fighter. In the ancient cities of the old world, from time immemorial, stone has played the chief part in the erection of buildings, with brick in recent years as a good second

But in America use has been naturally made of that material most ready to hand,—wood; and thus it is that the fire risk has grown proportionately to the population and the birth of new towns, both of which have been inordinately rapid and necessitating in their turn celerity of construction to meet the ever increasing demands of the situation. Further, a man will decide to build a mansion for himself, premises for his business or a factory for the production of some commercial article. He buys a plot of ground, selects an architect, chooses a set of plans and specifications, lets out the contract and is in occupation 120 days after he first conceived the idea. Hustle with a capital "H" is the keynote of the scheme, and any questions of fire control, appropriateness of design or strutural stability are all swamped in one wild desire for haste and the speedy completion of the order.

Contrast this picture with European methods, where the individual breaks fresh ground only after months, maybe

years, of careful consideration, and where the great grand-
son places the finishing touches to the conception of his
forbears. Again, take the ingredients commonly in use
for the mixing of mortar, that most essential adjunct to
building operations. The writer has seen in Europe pails
of animal blood and hair combined with the finest lime and
sand, it having long ago been recognized that the binding
qualities of this compound are unsurpassed. In America,
it is often thought sufficient to employ mud with a sprink-
ling of cement, which may be cheap but distinctly savors
of "jerry building."

It has sometimes happened that whole rows of buildings
have collapsed because of the inferiority of the materials
used in their erection, and often weeks after bricks have
been laid in this mud mortar two fingers would suffice to
pull them from the wall. True, some improvement has
taken place of late, but the buildings of the type mentioned,
carelessly finished within and without, provide that class
of construction generally described as "quick burners."

While on the subject it may also be remarked that when
a stipulation is made in a building contract for "fireproof
wood and finishings," the prospective owner of the premises
seldom realizes that wood so treated loses its fire-resistant
qualities from atmospheric moisture in a few years, while
hard wood will only absorb 35 per cent. of the solution
and maple none at all.

In discussing fire strategy in lofts and commercial buildings entirely new conditions present themselves to those
previously considered under the title "Fire Strategy in
the Homes of the People." The former structures have
large floor areas, usually heavily stocked with what may
prove to be combustible material, and in many cases open
stairways and elevator shafts, added to light shafts between two buildings, enhance the danger and the difficulty
of efficient fire control. Bold indeed would be the officer
who would attempt to lay down a set of strategic rules for
fighting fires in places of this type, as so many factors
enter into the problem that a prearranged attack is im-

OLD STYLE. POOR CONSTRUCTION

possible. The fate of nations has often been decided by a successful or a disastrous campaign, worked out by a military genius of the "Headquarters Staff," and it is related of Field Marshal Moltke that, on war being declared between France and Germany, he sent a telegram and went to bed; every possible detail of the war having been prepared months or possibly years in advance, even to the number of cups of coffee required at Cologne railway station for the arriving troops. Fire Chiefs must evolve their plans on the instant for they cannot calculate beforehand the strength of the enemy. Furthermore "fire" is the only adversary which on the battlefield steadily increases in strength in the exact mathematical proportion of the resistance it meets, until the point is reached when it is actually held in check. This emphasizes the part played in modern fire-fighting by promptness of decision, good judgment and rapid action on the part of those in command. Hence it can be realized that the mere theoretician stands a poor chance of acquitting himself creditably and that it is practice which tells.

But notwithstanding these factors, there are a few general rules which can and must be applied under any conditions. On arriving at the scene of an outbreak in a commercial building of the "quick-burner" type, the officer in command should be able to tell at a glance its height, width, depth and style of construction, and for this one moment should be sufficient. Here is a most important point, for no matter to what part of the building the commander may be obliged to go thereafter, he has a correct map of it and its surroundings in his mind's eye.

The next step is correctly to locate the seat of the fire, which can only be done by an instinct fostered by long training and experience, which becomes a sort of sixth sense and is therefore outside the boundary of rules and regulations. Some possess this faculty to a marked degree, while others seem to lack it utterly and, like an ear for music, it cannot be acquired though it may be enhanced and quickened by practice. Approaching the problem under consideration from a broad standpoint, the plan of attack depends upon the condition of the building and the extent of the fire upon the arrival of the first assignment. If the outbreak is on one floor only, a line forced up the stairs to its origin may be sufficient, and it should be added that this line usually extinguishes more fires than the three or four which follow. In fact, it may be hazarded that 75 times out of a hundred this will be all that is necessary. Should the flames be found to have control of several floors, the force should be deployed in such a manner as to confine the fire to the area it has already invaded, and should it be impossible to enter the building a water tower may be brought into action to cover the front, though the street is usually wide enough to obviate all danger of fire crossing and igniting structures on the other side. The rear also must be covered as there is infinitely more risk of the flames spreading to adjoining premises at that point, for such buildings are often within ten feet of each other. Should a fire burst out of the rear windows it will instantly cross this narrow space and, if not driven back by lines in the buildings behind, they too will ignite.

Should a condition of this nature confront the chief on his arrival, he should immediately summon additional aid and a common fault with some subordinate officers, which may cause fatal delay, is to postpone the transmission of further alarms. Precious moments are consumed before the fresh assignments can reach the scene and the fire has gained control. Therefore it must be reiterated that a correct estimate of conditions to be met is of vast importance. Once the rear of the building is covered the side exposures must be protected as there are four sides to a fire and the one to leeward is naturally the most dangerous. Seldom does a fire work to windward as it did in the case of the Equitable and hence a line or two should be stretched to leeward, the probable route of the flames.

When control of the outbreak is assured, with front, rear and sides covered, it may be possible to enter the building unless it be old and heavily stocked. Should there be fire escapes on the front, lines can be sent up by this means under cover of the water tower and turret pipes, and if lines are sent up the stairs the roof should be opened. This latter is a prime requisite to relieve the building of accumulated heat and gas which might explode and would certainly seriously hinder the actions of the firemen. As soon as the lines ascending the stairs and fire escapes have gained a foothold in the structure, the tower and turret pipes should be shut off, as there is no object in flooding the floors if the inside lines can control the flames. At this time the force in the rear should be advanced across the space and effect an entrance. Everything now would depend upon the condition of the building, for as a physician skillfully prescribes to suit the strength of his patient, so must the Chief cautiously advance according to the strength of the structure. If it is weak and tottering after a fire such as has been described, and furthermore contains stock that has absorbed a quantity of water thereby adding greater weight than perhaps the supports were intended to stand, men should pick their way; and if there is great doubt of its stability lives should not be risked until floors are relieved of all possible weight. Many deaths have been caused by the collapse of weakened buildings during what firemen technically term, "washing down." As the first word was celerity the last word must be caution, for since the fire is practically extinguished there is no necessity to risk valuable lives.

The fire protection of bonded warehouses offers a curious problem. They may not inaptly be described as quick burners since their construction the world over is on the same lines and apparently framed with no consideration of fire risks. For obvious reasons, outside fire escapes are barred, the presumption being that their presence might encourage the enterprising burglar or smuggler. Similarly the floors are of limited height, often with insufficient ventilation, economy of space counting for more than economy of fire risk. Which all seems futile and the reverse of far-sighted, though it has grown to be a commonplace that Governments are far behind municipalities in dealing with those common sense features which are so largely responsible for the safety and convenience of modern life. In Germany, it is true, the matter has received consideration and warehouses are constructed in such a manner that the minimum of damage is done to bonded goods, in the event of a fire, by having the floors raked and by the structures themselves being built upon the most approved designs. But this refers only to recently erected warehouses, notably in Hamburg and Bremen.

Now it goes without saying that as a rule the contents of these buildings are highly inflammable and hence every scientific nerve should be strained toward the adoption of some form of fire control, which shall meet the immediate demands of the situation. The sprinkler system naturally suggests itself as the remedy, but here questions of space step in, since the nozzles of sprinklers are always of a certain length and hanging from the ceiling would take up valuable room, let alone being peculiarly liable to damage during the shifting of goods. An alternative scheme has been evolved by the writer, which if not perfect, at any rate offers some feasible method of meeting the danger of fire and in construction is of a simplicity which speaks for itself. It has been christened the "Manifold system." Near the front entrance should be located a number of valves equal to the sum of the stories in the building. These should have pipe connections to each floor the nozzles of the same being finished just below the level of the ceilings and furnished with revolving sprays operated by water pressure. Most laymen are acquainted with the ordinary spray used on tennis lawns and grass plots; the principle is precisely similar. Upon an alarm of fire all that would be necessary would be for the officer in charge to ascertain the location of the outbreak and in case he could not reach it with a line of hose promptly to turn on the valve controlling the nozzles of that particular floor. The pressure would naturally be obtained from fire engines, high pressure mains or, if sufficient, from the street mains, though in cases where there was an installation of the high pressure system, the latter would be advantageous. In any event the fire would speedily be damped down admitting of the access of firemen to the building, while in many instances such means might in itself be sufficient to prevent further mischief.

This system of automatic fire-fighting is only in its infancy and the march of science will undoubtedly bring in its train increased efficiency of apparatus employed and the lessened possibility of its operating out of season, which sometimes occurs with the sprinkler installation. There is a crying need for the perfection of the self-acting fire-fighter, since in spite of modern fire resistive tactics, the enemy has itself kept abreast of the times and each new preventive method is offset by the introduction of some fresh element, which promises a splendid stimulant to the appetite of the flames. Thus the introduction of the automobile has led to the common use of gasoline, in itself highly inflammable and demanding special methods of storage. Scarcity of coal has turned the mind of the inventor toward the use of liquid fuel, while the advance of the photographic art has been responsible for the introduction of the cinematograph with its celluloid film. Most assuredly has this form of amusement come to stay, but equally its advent has not been an unmixed blessing to the fire-fighter since the dangers connected with its operation are so diverse and ever present that special precautions to meet the same have ever to be framed. And thus it is along all lines of advance; if the human brain is never idle, then most assuredly the fire fiend is never quiescent and is prompt to seize upon fresh opportunities of attack.

Hence in considering "quick burners" as a whole, it may not be inappropriate to include a few words anent the moving picture peril, since in all truth this strikes at the foundations of the social system owing to the number of children who habitually frequent such places of amusement. Now, in the first place, if a panic in a theatre is a tragedy, then a panic in a moving picture hall is doubly so, since it is hopeless to appeal to the self-control of the audience and the strong chances are that once a rush for the exits begins, nothing will prevent confusion and crushing. Aisle guards can accomplish little if the principles concerned are not, in the bulk, amenable to the dictates of reason and if old heads are not to be found on young shoulders and if amongst the former panic is not uncommon, what then can be expected from the latter?

Hence it is that the picture palace should be as fire secure as human knowledge can make it, which, in spite of municipal regulations, is seldom the case. From the nature of the entertainment apart from the actual apparatus, all that is required is a white sheet upon which the pictures are displayed and thus practically any sort of hall will meet the case. Old churches, disused stables, deserted chapels, in fact any building which is good for nothing else is impressed into the service and with a coat or two of paint blossoms forth under a new guise as a "picture palace." So long as the requirements of the niunicipality have been fulfilled, there is no cause for interference and so it continues to thrive until the day comes and fire sweeps along laying it low like so much match wood and demanding a heavy death roll of women and children.

Now, there is one point in this connection, which, small in itself, is really the kernel of the situation. Ninety per cent. of film fires occur in the immediate vicinity of the operator and yet the observant will have noticed that nearly always the box containing the apparatus is over the entrance door or, where there are several doors, at the end of the hall where the entrances, which often serve as exits, are situated. There is no sufficient reason for this and it would be just as convenient from a managerial point of view were the position reversed and the apparatus located at the end of the structure remote from the entrance. The reason for this alteration is obvious. In the event of a fire, at present, the audience is obliged to pass out, either alongside or underneath the probable seat of an outbreak, an unpleasant and dangerous task for grown people, an impossible one for children of tender years. Now it will be argued that the adoption of such an arrangement would spoil the performance, since the shadows of incoming patrons using the centre aisle would be reflected upon the picture curtain. The answer to this objection is apparent. It would be an excellent move if the centre aisle should be used solely as an emergency exit, sufficient width being allowed to the side aisles to render both ingress and egress easy.

There are, of course, many other structures, which from the nature of the trades carried on within may be said well to merit the epithet "quick burner." In fact, in such cases it is a question of contents rather than construction, and in this connection a special chapter has been devoted to the consideration of the storage of gasoline and the garage peril. There are, however, apparently harmless factories which provide the wherewithal for dangerous explosions, and it will be well to give some slight consideration to these. For the benefit of the layman, the simplest course will be to supply an illustrative parallel. Take an ordinary log and try to burn it. Short of placing it in a furnace it is next door to impossible to incinerate it. Even after a severe fire, hard-wood beams of some thickness are rarely burnt through, though naturally their outside surface is charred perhaps to the depth of an inch or two. But split the log and its component parts will burn more readily, while the smaller it is chopped the easier it catches alight, until the point arrives when it makes excellent kindling. Reduce the log still further and it becomes sawdust, which is not only highly inflammable but under certain conditions actually explosive. Hence it follows that the greatest precautions must be adopted in all factories or warehouses in which large amounts of sawdust are liable to collect. This doctrine may be extended, and might not inaptly be termed "the dust danger."

Flour, ground grain of any kind, all belong to the same category, and offer the same risks. Therefore let every manufacturer or warehouse man beware of accumulated dust which, should the slightest outbreak of fire occur, will become a potential explosive. Incidently, there is no reason why factories of the nature mentioned should not be kept clear of this menace, as, in so many other instances of aggravated fire risks, they are as often as not directly accountable to the element of carelessness inherent in human nature.

It may come as something of a shock to the lay reader to know that drug stores, or chemists' shops as they are called in England, are amongst the most difficult problems the fire-fighter has to handle. Heavily stocked with all sorts of acids and alkalies, no chemist on earth can precisely foretell what results may not follow upon some unforeseen chemical combination. An explosion may occur capable of wrecking a whole block of buildings, as was the case in the Tarrant Building in New York in the year 1902, or, poisonous fumes may be generated which will render it almost impossible for firemen to operate within a considerable radius of the spot, unless equipped with smoke helmets. It is no exaggeration to say, that the harmless looking little drug store at the corner of the street, is a factor of such danger that in New York city special regulations have been framed for fire prevention in such establishments.

For consider the perilous possibilities of such a common chemical as chlorate of potash, an excellent remedy for sore throats and coughs, and under certain conditions of the greatest medicinal value. Yet the following details of the peculiar activities of this substance cannot fail to supply food for thought, when it is remembered that its characteristics are not uncommonly met with in other articles usually supplied by the local druggist.

Chlorate of potash is a white chrystalline body found in commerce in crystals or in a powdered form. It consists of the metal "potassium" and the gases "chlorine" and "oxygen," chemically combined to form a potassium salt of chloric acid. The proportion of oxygen is large as compared with that ordinarily present in salts and is very

weakly held in the combination. This makes chlorates as a class, dangerous compounds, as heat alone will liberate the oxygen, leaving behind "potassium chloride," a compound similar to table salt.

Danger arises from chlorate of potash in four ways: (1) When mixed or in contact with combustible substances and ignited, an explosion results which proceeds with tremendous energy and fierceness making a bad fire and one dangerous to fight. This violence of action is due to the liberation of pure oxygen which immediately attacks any explosives present. (2) When mixed or in contact with combustibles such as charcoal, sulphur or sugar, particularly if both be finely divided, and abraded or struck the result will be an explosion. Spontaneous explosions may also occur, and near contact between chlorate and yellow or stick phosphorous is frequently followed by a violent explosion. (3) Strong sulphuric acid in contact with a chlorate will cause it to decompose and to give off heavy yellow gases which are explosive by even slight shocks or by contact with easily oxidizable material. These gases will spontaneously enflame phosphorus, turpentine and other substances. (4) From the inherent character of the salt itself.

Chlorates, for theoretical chemical reasons, are in one sense unstable compounds. The danger from the third source is obvious, although it is not imminent from the third and fourth as it is from the first and second. In the first case, it will be seen that the presence of any considerable quantity of chlorate in a building is a source of danger. It would give such impetus to a fire when once reached as to make the destruction of the property almost certain. When heated alone it gives up its oxygen quietly, but in the presence of combustibles, an explosion will result from the rapid generation of highly heated gases. A fall of floors or of shelves might scatter the chlorate over a large surface already hot, or might mix it with highly combustible materials such as are usually present in drughouses. In either case disastrous explosions will occur.

In the case of the disastrous fire already mentioned at Tarrant & Co.'s warehouse, one of the great wholesale chemical houses in New York, this is what probably occurred since tons of chlorate and sulphur were stored on one floor and the ensuing explosion consequent upon the fire completely destroyed the building and several others adjacent to it. In the second case there is danger from intimate mixtures of chlorate and combustibles. These will be found almost exclusively in torpedo and fireworks factories and have the explosive force of dynamite and guncotton and in fact may be placed in the same category. The explosion is propagated through the mass by shock and is practically instantaneous, while in gunpowder there is simply a very rapid combustion generated by flame or heat alone. For full effect a chlorate powder must be confined, or in such quantity as to produce the effect of confinement, in the mass. In case of carelessness, such mixture could occur in small amounts in any drug store or chemist shop, and if stepped upon would ignite. The danger is, that, while the explosion in unconfined portions of such mixture takes place usually only in the part under pressure, the action is continued in the remainder of the mass as a fierce and very rapid combustion resembling that of red Greek fire.

The danger in the fourth case is a doubtful quantity, but there seems to be some grounds for belief that chlorate will, of itself, decompose with explosive violence, if exposed to heat and shock at the same time. Alone and unmolested, chlorate is a perfectly safe and stable compound, and the peril arises from surrounding circumstances.

This brief description of one of the commonest of chemical commodities will give the layman some idea of the precautions which should be adopted to prevent fire in all drug stores, wholesale and retail, and in New York this fact has led to the adoption of a detailed standard as to the amounts of dangerous chemicals which may be stored on licensed premises. In the first place those entering the chemical business must obtain a permit from the Fire

Commissioner, who, with the Municipal Explosives Commission, considers each application upon its individual merits. In the event of any disregard of regulations, licenses may be immediately revoked.

The following are some of the details of the regulations regarding retail drug stores. Firstly, it is unlawful to manufacture, compound, dispense or store upon such premises any of the following substances: Colored fire in any form, flashlight powders, liquid acetylene, acetylide of copper, fulminates or fulminating compounds, guncotton, gunpowder, chloride of nitrogen, amide or amine explosive, picrates, or rubber shoddy.

Potassium chlorate in admixture with organic substances, phosphorous or sulphur is forbidden, but this restriction does not apply to the manufacture or storage of tablets of this chemical, when intended for medicinal purposes. Much the same applies to nitroglycerine, which is rigidly barred except in medicinal form, as approved by the National Pharmacopœa.

A schedule has been arranged, limiting the amount of combustible chemicals and fibres, some of which it may be of interest to quote. Thus, carbonic acid 100 lbs., collodion 5 lbs., turpentine one barrel, essential oil 100 lbs., in all, phosphorus red and yellow 3 oz., magnesium powder and ribbon 16 oz. in all, powdered charcoal 10 lbs., rosin 10 lbs., lint 10 lbs. in closed boxes, potassium permanganate 5 lbs., silver nitrate 1 lb., glycerine 500 lbs. Hence it will be seen that a limit is set upon the storage of even the commonest commodities, due regard being taken of their possible combinations in the event of accident or fire.

The next official regulation might well find a place in chemists' shops the world over. "It shall be unlawful for any person to store or accumulate broken wood, waste paper, or waste packing material of any kind in any part of the premises where goods are packed or unpacked. Such materials should be removed at the close of the day."

In addition, the following restriction is an admirable one, and might well be extended. "It shall be unlawful for any person to smoke or to carry a lighted cigar, cigarette or pipe, or any lighted substance, within a packing room, cellar, store-room, or that part of the laboratory where volatile inflammable oil or liquid is used or handled; and a notice bearing in large letters the words, 'SMOKING UNLAWFUL,' together with a copy of this section in small letters, shall be conspicuously displayed in one or more places on each floor." Furthermore, basements and cellars must be properly lighted by electricity, and persons neglecting this and the above regulations are guilty of misdemeanor.

It may seem to some that these ordinances are stringent to the extent of being irksome, but it cannot be too strongly emphasized that the every day fire risks of a community are in themselves amply sufficient with which to deal, and that those exceptional hazards demand exceptional precautions owing to the unknown character of the outbreak which may result from accidents thereto.

There is one last example of the genus "quick burner," that the writer would like to mention. There is a craze nowadays, in all parts of the world, for the small suburban home, which with its ornamental exterior, tessellated pavement and brightly painted front door, appears to the average purchaser an epitome of desirability. By the fire-fighter, however, these rows of jerry-built cottages, hastily run up by an unscrupulous contractor, are rated at their true value. As a rule, examination will show that there is a common "bearing wall" between each two houses, rarely if ever extending to the top of the attic. Thus, the whole length of these attics, unpartitioned off in any way, forms a huge horizontal flue and an excellent ally to the flames.

In dealing with an outbreak of fire under such circumstances, it is therefore necessary to take lines in six or seven houses away from the actual scene of the blaze, in order to fight the flames back and to prevent them from gaining complete control and sweeping all before them like so much waste paper. The structural disabilities of such a system are vexatious enough since tenants are unable to effect any architectural alterations, but the fire risks are tremendous. It is another example of that get-rich-quick policy, which does not concern itself with such elemental factors as fire risk and human safety, and is occupied solely with its own selfish ends.

In fact it is no exaggeration to say that the greater portion of fire legislation is concerned with the protection of the individual against the egotistical indifference of those who are ready to exploit him. Fire resistive construction costs little if any more than "jerry building," due regard being taken of durability, security and reputation, though apparently the latter counts but little in comparison with lightly earned gold. The public, however, is happily commencing to take an intelligent interest in fire control and the day is drawing to a close when it will be possible to gull the unwary by means of cheap ornamentation and a prolific use of paint. Or perhaps a hint might be taken from Germany, where owner and occupier are held jointly responsible for outbreaks of fire.

CHAPTER XXI

INCREASING land values due to congestion in large cities, coupled with advances in the mechanical arts and steel skeleton construction have ushered in a new and perplexing problem for the fire-fighter. In American cities there is no restriction placed upon the height to which an office or commercial building may rise, as long as certain regulations are complied with in regard to the material used. Height is not prescribed by law, but by economic conditions. Were it possible to provide inexpensive elevator service, there is no reason to doubt that buildings in New York would now have a hundred stories as there are already several with forty, and one with fifty-seven, floors. In the old type of factory or commercial building, ranging from one to six stories, the ordinary way of stretching hose lines by stairs and fire escapes provided all the necessary means for the firemen to reach the seat of an outbreak, but in these higher buildings other methods had to be evolved. Stand-pipes running from the lowest to the highest floors were introduced. At first these were crude affairs often misplaced and also deficient in size. However, experience soon discovered these defects with the result that in New York an ordinance was passed, which it may be well to quote.

"In every building now erected unless already provided with a three-inch or larger vertical pipe, which exceeds 100 feet in height and in every building hereafter to be erected exceeding 85 feet in height, and when any such building does not exceed 150 feet in height, it shall be

provided with a four-inch stand-pipe running from cellar
to roof, with one two-way three-inch siamese connection
to be placed on street above the curb level and with one
2½-inch outlet, with hose attached thereto on each floor,
placed as near the stairs as practical; and all buildings now
erected, unless already provided with a three-inch or
larger vertical pipe, or hereafter to be erected, exceeding
150 feet in height, shall be provided with an auxiliary fire
apparatus and appliances, consisting of water tank on roof,
or in cellar, . . . and such other appliances as may be
required by the fire department. . . . Stand-pipes shall not
be less than 6 inches in diameter for all buildings exceeding
150 feet in height.   All stand-pipes shall extend to the
street and there be provided at or near the sidewalk level
with the siamese connections.   Said stand-pipes shall also
extend to the roof. . . . If any of the said buildings extend from street to street, or form an L shape, they shall
be provided with stand-pipes for each street frontage."
As will be seen this sets a minimum requirement of 4-inch
stand-pipes for all buildings over 85 and up to 150 feet
in height, and 6-inch for all buildings over 150 feet.   In
practice it was found that a greater diameter was necessary in order to give more water for fire-fighting, and recently the diameter was increased to eight inches in buildings over 150, and 6 inches for buildings under that
height.

At this point a brief description of a stand-pipe, its equipment and operation may not be out of place.   The material
used is generally galvanized, sometimes ordinary black pipe.
In the case of very high buildings the main line must be
connected by a Y at the street level, giving four 3-inch
inlets.   These inlets to the siamese have "clapper" valves,
serving as checks to prevent the water from backing out.
There is also placed a swing check to prevent the water in
the pipe from backing out to the street connection when
the line is not in use.   This is to prevent freezing in winter
time.   The object of the pipe line being connected to a
tank on the roof is to ensure a supply of not less than

3,500 gallons of water, which will enable the occupants of the building to hold in check an incipient blaze prior to the arrival of the Department. All house supply lines for domestic purposes attached to the tank, must be tapped in at such a height as to ensure at all times the quantity of water mentioned. A check valve is placed at the bottom of this tank on the stand-pipe line, set to permit a downward flow and to prevent an upward flow, in case a fire engine or high pressure lines are connected at the street level. The omission of this check would prevent pressure on the floor below, as water pumped in would merely overflow from the tank on to the roof. An open hose outlet on the roof will have the same effect, as the writer has more than once observed. Officers should carefully note these points in making inspections. Templates should be provided and connections at street level and also those on the outlets on floors should be tested to see that they comply with department standards. It is a very serious matter, when a dangerous fire is raging in the upper stories of such structures, to find that the connections will not fit the hose and that water cannot be forced through the stand-pipes except through the medium of inside connections, a somewhat ineffectual expedient, but that hose lines must be stretched 12, 14, 16 or more stories.

To the unthinking this may represent little, but to the experienced fire-fighter it means much. Very few men would be able to walk up fourteen flights of stairs and then be fit for work on their arrival. Firemen in New York must often climb more than 150 feet, but it is impossible to drag hose lines up to such height. Also the hose cannot stand the excessive pressure necessary to force water to this elevation, and even were it possible, the time employed in so doing would permit the fire to extend to such proportions as to destroy the building. With sufficient material on which to feed, a fire will extend in geometrical ratio to elapsed time, which again proves the necessity of speed in order properly to control an outbreak. All signs regarding pipes should be properly placed.

Where there are two lines of pipe, one for the sprinklers or perforated pipe system, and the other for the stand-pipe, a mistake or misplacement of the signs indicating their nature will cause serious delay and often great damage. At this point it may be well to state that perforated pipes are a poor substitute for sprinklers and should be taken out of all buildings and the latter installed in their place.

The street connections should be not less than 2 nor more than 3 feet from the sidewalk level and at right angles to the building, which facilitates the work of connecting up. Should these connections be out of order, lines can be connected to the outlets on the lower floors by means of a double female connection. The method of procedure is to disconnect the house line, put on an increaser—$2\frac{1}{2}$ to 3 inches—leaving a three-inch male thread, to this connect the double female; this will give a three-inch female connection for the male end of the hose stretched from the engine or high pressure hydrant. As soon as connection is made, open the valve and start the water. All outlet valves on floor outlets should be of the gate type, which gives a free waterway, whereas the globe type materially reduces the flow. Where several lines are being used on the upper floors of a building, and the two or four lines connected to the siamese are overtaxed, the supply may be augmented by connecting to the lower floor outlets as previously described. In New York, stand-pipes are used in buildings so equipped when the fire is on or above the fifth floor. Below this point the ordinary method is preferable. The hose provided in these buildings should be of good quality, capable of sustaining a pressure of 200 pounds and be fitted with 1 or $1\frac{1}{8}$-inch nozzles. In many cases, through lack of attention on the part of owners or agents to keep the hose in proper condition, the firemen fear to trust it and connect up a department hose, using their own controlling nozzles for inside work.

In order to describe the actual use of these stand-pipes **and** the pressure required to force water to the upper floors of buildings on fire, let it be assumed that there is an outbreak on the sixteenth floor of a 20-story building. The companies arrive and connect to the stand-pipe, and as the fire is a threatening one four lines are led to the siamese. Hook and ladder and engine company men proceed to the floor directly beneath the fire, by order of the Chief, using the elevator for this purpose. The ordinance requires that night and day elevator service must be provided in all buildings over 150 feet in height. Each company has its instructions, and a Battalion Chief, in command on the inside accompanies them. As soon as connections have been made to the siamese, the chief officer must determine the pressure necessary to furnish effective fire streams on the sixteenth floor, or an approximate height of 170 feet. The formula, for this procedure is to multiply the height of the building by .434, but here there is no time for pad and pencil. The best method is to allow 5 pounds for each floor of elevation, and at the height above mentioned it can be readily grasped that the result is 80 lbs.—7 lbs. more than the exact formula but which will be needed as the stand-pipe goes from the sidewalk to the basement and perhaps to the sub-basement. This gives the column additional length and as the stand-pipe often makes a right angled turn at the basement level there is an additional loss of pressure at that point. Also there is a nominal friction loss in the pipe itself. The writer's purpose is not to weary the reader with technical details, but it is necessary to explain this procedure for the benefit of commanding officers at fires, at greater length than many other subjects with which firemen are more familiar. Therefore, 5 lbs. will be allowed to each story for weight, about 4 lbs. friction loss for each 50-foot length of 3-inch hose, 5 lbs. loss at siamese connections and 10 lbs. for entry head at the valve outlet on the upper floors.

It will be assumed that there are two lengths of 3-inch hose in each line, giving 8 lbs., 5 lbs. for siamese, 80 for weight of water column and 10 for entry head; making 103 lbs. in all. Add to this the force necessary to

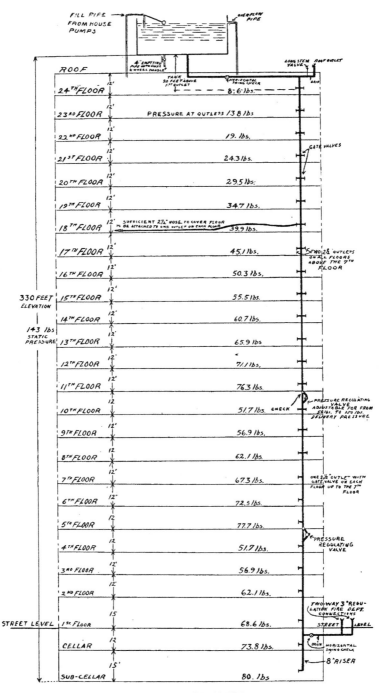

DIAGRAM SHOWING APPROXIMATE PRESSURE
IN SECTIONS BETWEEN REGULATING VALVES.

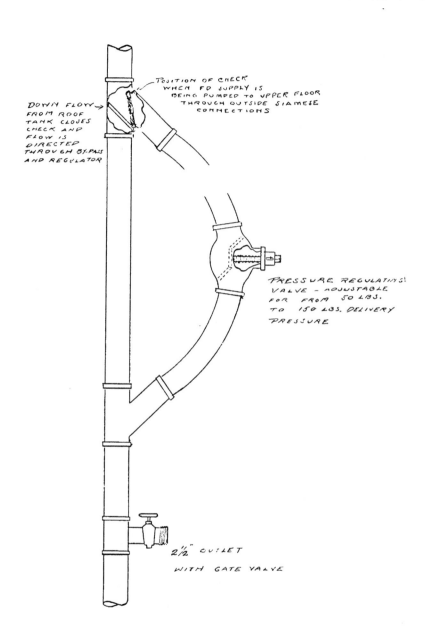

POSITION OF CHECK
WHEN FD SUPPLY IS
BEING PUMPED TO UPPER FLOOR
THROUGH OUTSIDE SIAMESE
CONNECTIONS

DOWN FLOW→
FROM ROOF
TANK CLOSES
CHECK AND
FLOW IS
DIRECTED
THROUGH BY-PASS
AND REGULATOR

PRESSURE REGULATING
VALVE - ADJUSTABLE
FOR FROM 50 LBS.
TO 150 LBS. DELIVERY
PRESSURE

2½" OUTLET

WITH GATE VALVE

DETAIL SHOWING POSITION OF
REGULATION VALVE ON BY-PASS OF STANDPIPE.

give effective nozzle pressure which would be from 50 lbs. and upward according to the extent of the fire and the result is 153 lbs. This, of course, might be varied if a number of streams were taken off the stand-pipes, as the friction loss in hose and pipe is as the square of the velocity. That is, if water is travelling at the rate of ten feet a second the square of ten is 100, but let the flow be increased to 20 feet per second, and the square is four hundred, or four times as much. Therefore, while the velocity would only double the friction loss would be four times greater. Officers arriving at an outbreak, find, as by a predetermined order, that the pressure on the mains is 100 or 125 pounds. They must judge instantly if additional pressure is required and order it promptly, though it must be remembered that much damage may be done by excess in this direction. The writer once arrived to take command at a fire in a five-story commercial building. The structure had 50 feet frontage and was 200 feet deep and the fire had control of the three upper floors and was threatening surrounding property. The officer found in charge was asked what he had done and if additional companies had been summoned, as there was but a first alarm assignment at work. The reply was, "No, sir, I did not send out a second alarm, but I ordered 250 lbs. pressure on the high pressure pumps." This was soon after the installation of the high pressure service and officers had not yet mastered its proper use. In this case 125 lbs. was quite sufficient, or at most 150, to which point the pressure was immediately reduced and additional forces summoned. The very high and unnecessary pressure might have caused damage to hose, apparatus, pumps or even men. Officers who wish to master these technical details should study them at leisure in their offices, so that they may be available for instant practice at a fire.

The best method of fighting fires in lofts or office buildings of great height is to connect the first line at the outlet floor below the fire, lay the line out free of kinks and stretch up the stairs. Charge the lines. Never be caught with an empty line in hand when the door is opened. Here is where many a man is injured, for his old enemy, "back-draught," is just inside and the moment the door is opened it rushes out with superhuman force.

Lie low and like a pugilist, duck the blow, and open up the pipe at once. Without going into a scientific discussion of the causes of back-draught, it may be well roughly to define it and to describe its action from a fireman's viewpoint. Plainly stated, it is an excessive pressure on a floor where windows and doors are tightly closed. As the heat increases the air expands, causing a greater pressure on the floor than the corresponding outside atmosphere. When a vent is given there is a blowout, resembling an explosion, for the pressure must be equalized. In other words the pressure on the floor must be brought down to the corresponding atmosphere on the exterior. The first explosion may be slight, but is often followed by a more dangerous outburst, depending upon the degree of heat on the floor. The influx of fresh oxygen supplies abundant fuel for the flames and the smoke, which is part of the goods imperfectly consumed, bursts into flame and shoots through the opening. This condition must be guarded against, as men may be caught on the stairs and literally roasted. The sudden outburst of flame may also blow out windows on all sides, endangering surrounding property or the floors above. Another, and still more dangerous form of back-draught, is where a slow burning fire consumes the oxygen in a building tightly sealed. A gas much lighter than air results, and this is particularly aggravated by the nature of certain kinds of goods. As is well known, different chemical changes take place in different kinds of material. A vacuum is formed inside the building and as soon as the latter is opened there is an inrush of air which coming in sudden contact with the gas, causes an explosion. The writer has often seen this peculiar phenomenon. The explosion is preceded by an awe inspiring silence; nature seems suspended for the moment while the two elements meet. In an instant, all is wreck and ruin.

Men of experience instinctively feel this condition, and the order to back down and out is instantly given, though, if in a position where they cannot get away at once, men should drop on hands and knees, covering their faces from the wave of flame. It may pass over without doing damage, when they should instantly arise and keep the pipe open while backing out of danger. Perhaps in all the science of fire-fighting no part requires more care and attention than the opening of buildings on fire. Many brave men have lost their lives at this dangerous work. At a fire in the stock yards in Chicago, Chief Horan and 23 officers and men were killed just as the door was opened. There were no windows in the building through which the pressure could be relieved, and the wall gave way under the strain, burying all who were in front of it. Where such conditions are suspected, an opening should be made in the roof; this will relieve the pressure and is a fairly safe method.

A brief reference must be made to the following important point. In cases where fire has complete possession of 1, 2 or 3 floors in a high building, beyond the reach of towers or turrets, great care should be exercised in placing lines above the outbreak. Companies of men with good streams should be at the doors below in order that the fire may not come out and cut off the retreat of the men above. Buildings of this class have more than one stairway and more than one stand-pipe, and the least exposed position should be chosen in advancing lines to the upper floors. No line should be sent to an upper floor until the companies operating the one below have gained a foothold inside the door, with a reasonable assurance that they can hold the position. Give as much ventilation as possible in order that the ascending smoke and heat may be minimized. Chiefs should avail themselves of rear fire escapes and every other possible point to get ahead of the fire, for should it once become uncontrollable on the upper floors men cannot be kept in the building, and a collapse may momentarily occur. In all buildings of this class fire towers should be provided. These are separate and distinct from the main building and in order to reach them it is necessary to pass from the floor through a fire-proof door on to a balcony outside the building, and through another fireproof door into the tower itself. Or as a variant to this system there is a plan whereby every building is to be divided into two sections by means of a fire wall running from cellar to roof. This fire wall would have fireproof doors on each floor and there is almost no chance of a fire starting simultaneously in both parts of a building so divided. In case of an outbreak in one section, the people on each floor would walk through the doors to the other section, shut them on the fire and take their time to reach the street and firemen could easily gain a foothold on a floor subdivided in this manner. A stand-pipe, in the former case, should run from lowest floor to roof through the tower. Firemen could pass ahead of a fire and attack it on each floor with perfect safety. For rescue of occupants, this means, combined with the afore-said horizontal fireproof partitions subdividing the floor area, would make the buildings safe for public and firemen alike. Add to these two indispensable requisites a thorough installation of automatic sprinklers, and the high building would be shorn of its terrors.

In conclusion, the writer feels that after the foregoing pages, with their somewhat technical details from the lay-man's point of view, it may not be without interest to give some slight description of the greatest skyscraper in the world, the Woolworth Building of New York. Incident-ally, this gigantic block of masonry was designed by its namesake as a memorial of his earthly success and to the glory of the commercial enterprise of America. Including basement and sub-basement, it consists of 57 stories, and rises to a height of 790 feet above street level, with a main frontage of 150 feet. Thus it will be seen that it is five times its own width and is less than 100 feet short of the Eiffel Tower. It is, of course, of fireproof steel frame construction. The preparations for the foundations were begun in August, 1911. On May 1st, 1913, the doors were thrown open to tenants. This represents a period roughly of twenty months, during which time nearly three stories must have been added every thirty working days. Besides containing business premises, which, incidentally, are arranged as regards floor space to suit the wishes of tenants, it is equipped with two restaurants, one in the cellar and one on the twenty-ninth floor, Russian and Turkish baths, a swimming pool and ice plant. All window frames within 30 feet of any other building are of hollow copper glazed with wire glass. It has four enclosed stairways of iron and marble, though only one extends to the top of the tower. There are 28 passenger elevators built on the most up-to-date fire resistive principles. For the protection of the building, the following equipment has been installed:

There are six 6-inch risers running from the sub-cellar to the 30th floor, two from the 31st to the 41st floor and one from the 41st to the 55th floor fed by tanks on the following floors:

> 1 Tank   6,300 Gals. on 14th floor, feeds from sub-cellar to 12th floor.
> 2 Tanks 10,000 Gals. on 26th floor, feeds from 13th to 24th floor.
> 1 Tank   3,100 Gals. on 37th floor, feeds from 25th to 34th floor.
> 2 Tanks { 1—6,700 Gals. / 1—3,200 Gals. } on 50th floor, feeds from 35th to 48th floor.
> 1 Tank   1,200 Gals. on 53rd floor, feeds from 49th to 53rd floor.

There are four outlets above this tank with no supply.

The main riser, 6-inch, has checks on the 14th, 27th, 37th, 50th and 53rd floors with checks on the horizontal run.

There are two suction tanks in the sub-cellar:

> 1 30' x 9' x 9'
> 1 15' x 9' x 9'

There is a swimming pool in the sub-cellar Turkish baths which can be used as a suction tank, with a capacity of 30,000 gallons.

There is one Dean Electric Pump, $3\frac{1}{2}'$ x 6' with a capacity of 300 G. P. M.

There are in addition five Worthington Steam Pumps as follows:

```
1—20' x 9' x 18'      500 G. P. M.
1—20' x 10½ x 15'     500 G. P. M.
1—14' x  6½ x 15'     150 G. P. M.
1—14' x  6½ x 10'     150 G. P. M.
1—12' x  4½ x 10'      80 G. P. M.
```

All pumps can be operated singly or collectively, and are supplied by street mains as follows:

```
1—6" from Broadway.
2—3" from Park Place.
2—3" from Barclay Street.
```

Such are a few details of this extraordinary structure, which may be expected to house daily some 10,000 souls. And there is no reason to believe that finality has been reached. Room must be found for the teeming thousands who throng to the business section of New York, and expansion must occur by the way of least resistance—upward.

CHAPTER XXII

APPARATUS FOR FIRE-FIGHTING

NOT since the first fire department was organized in ancient Rome has there been such an awakening as at the present time, and all countries are vying one with the other in the development of fire controlling and fire preventing appliances. In fire control the greatest effort is being made to develop motor propelled apparatus and motor driven pumps. The method of propulsion is almost perfected, as all that was necessary was to apply the principles so successful in moving pleasure and commercial vehicles to fire apparatus. But with the motor driven pump matters are somewhat different. Perhaps no mechanical device used in the control and extinguishing of fire has reached such a high state of perfection as the steam pump. To discard this time honored and often severely tried machine for a practically untried device, untried at least under severe conditions such as being forced to deliver water under high pressure for 35 or 40 hours or longer, would be unwise. Many American cities have a fire hazard due to old and faulty building construction, narrow streets and severe weather conditions, where it would be almost criminal to install a new type of apparatus until it had been tested in places where the risk would be limited.

It was the feeling that absolute reliability must be guaranteed which caused many cities to hesitate in adopting the motor driven pump. An extensive field was found, however, for this type of apparatus in suburban settletlements and in small towns. In these places the fire force is usually a voluntary one, horses are not always available nor can fuel easily be supplied, and should it happen that there is not sufficient water pressure in the mains a motor-propelled and motor-driven pumping machine is a great boon. It requires but one man to take the machine to the scene and it can be accomplished in one-tenth of the time required for a horse-drawn or hand-pulled engine. Time is all-important in such cases, and as the run is often long and the gradients heavy, great damage may be done before the arrival of firemen or apparatus which usually results in the total destruction of the building involved. Therefore in such places and under such conditions motor-propelled apparatus is, and motor-driven pumps may be, desirable; but in large cities, where great congestion intensifies the fire hazard, the first efforts of manufacturers were not successful in producing a motor-driven pump to meet the requirements of the Fire Department.

The scrap heap was large in those days, but there is some evidence that it is diminishing. Manufacturers found by the only reliable method, that of experience, that some of their pet ideas were not feasible. Chiefs of Fire Departments discovered this to their cost, but out of confusion is being evolved a fairly good motor fire engine. Naturally, it is still lacking in the perfection of the steam pump, but when it is considered that fifty years were required to develop this apparatus, it must be allowed that in possibly one-quarter of this period a most excellent motor-driven pump will be in use.

The steam pump was brought to perfection along the line of reciprocating motion, steam being particularly adapted to this movement as shown in all kinds of steam engines, marine, stationary, etc. But the application of a somewhat different force, that of gas explosions, to a piston, developing what is practically a rotary motion, the writer is inclined to think will not be so efficacious. For any man to attempt to condemn the reciprocating system would be folly; it may work, and as a matter of fact does work, fairly well, several firms having turned out average machines of this type. But in converting one mo-

tion to another there is necessarily the introduction of complicated gears and a multiplicity of parts, often hard to reach in case of repairs, and there is in any case a loss of power in transmission which would seem to militate against the piston pump.

As some pin their faith to the reciprocating style, so others favor the rotary. This type seemed to promise success, as there was no conversion of motion and the gear was much simpler. However, some new and many old defects showed up, and the great fault of excessive slip known to exist in all rotary pumps when run at high pressure was still apparent. Close-fitting casings with the introduction of improved springs behind the gibs seemed to promise success, but under severe trial were glaringly defective. No chief of experience could be satisfied with this. The manufacturers greatly improved the old rotary type of twenty-five years ago. The diameter of the shaft being increased to take out spring under high pressure. The shafts were shortened for the same reason. Water pockets between the gears, which gave a back pressure, were changed by tapping into the pocket or hollowing out the casing and piping around to the discharge. But still there is room for improvement. When a very high pressure is required, a point is reached when a rotary pump acts almost like a piston pump, each gear striking a blow exactly like the plunger, when the point of closest contact to the casing is reached. This is bound to jar injuriously the entire machine, and to wreck it unless it is very strong. Such are a few of the defects which came under the writer's notice when watching tests. There are many things to be said for and against all types, but the opinion may be hazarded that even the makers of these rotary, or gear, pumps know and admit that this apparatus is not the equal of the steam pump.

Another type tried with moderate success is the centrifugal. That this would be a immediate success there is not the slightest doubt, were it not that, like the reciprocating pump, it is somewhat difficult to use it in conjunction with the motor. In order to obtain a centrifugal pump which will deliver 750 gallons of water per minute at 130 lbs. pressure per square inch and come within the weight practical for fire apparatus, it must run at a speed of 16 to 18 hundred revolutions a minute. For fire apparatus, at least, this is not practical, for a new difficulty arises, that of cooling the cylinders. High speed is possible in fast moving cars when a strong current of air is forced through the radiator, but in fire apparatus standing stationary at a fire the designers are confronted by very different conditions. Therefore, in order to obtain the speed, a differential of at least two to one is necessary, or the other alternative must be used; that of increasing the size and driving direct at the same speed as the motor. In the one case there is the differential gear with some loss of power, and in the other a slow but much heavier pump. Could a motor be built that would directly drive a centrifugal pump at a speed of twelve hundred revolutions a minute, it would be the ideal type for motor apparatus, and the writer believes that one will soon be obtainable.

As already stated in this volume, there are various devices for raising aerial ladders, some mechanical, others electrical and still others by means of compressed air. It is not the purpose to enter into an explanation of how these devices should be operated as any intelligent fireman can learn their use in half an hour, but a word may be said as to the placing of the ladders themselves. The writer has often seen ladders in such a position that it was quite difficult to work on them to advantage, whether the work was rescuing persons or operating lines.

In order to obtain a clear idea of the proper position in which to place a truck so that an aerial ladder may be properly operated, let it be assumed that the ladder is to be raised to a window on the sixth floor of a commercial building. This would be about 65 feet from the ground, as the first floor is about 15 feet and the others about 10 in height. The approximate measurements can be determined by a practical truck man at a glance, and the truck should be set about 16 feet from the building with the centre of the turntable in line with the centre of the window. The point of the fly ladder should extend about 15 inches above the sill, but care should be taken in placing the same that it should not be permitted to rest heavily on the sill, but should be lowered to within a few inches of it. As soon as the weight of a body is near the top it will cause the ladder properly to set against the window ledge. Should the ladder be originally placed as above, there is no danger of its buckling in the centre. Reference is made to this matter of properly placing ladders on account of the great difference in the width of sidewalks and some officers accept the sidewalk as an infallible guide. Of course it is not intended that the space should be measured off with a two-foot rule, but a little thought and experience works wonders, and if they would only practice with ladders of various sizes they would quickly pick up the faculty of placing them in proper position. In the New York Fire Department, frequent drills are held and it is surprising to see how rapidly even untrained men grasp the idea. A poorly placed ladder at a fire is inexcusable.

For those who care to go into the subject a little more deeply and perhaps to practise at the drill school with apparatus, the following rule may be helpful: Divide the height to which the ladder is to be raised by five and add three. Thus, take a thirty-five-foot ladder. Five into thirty-five goes seven times, add three and the result is ten. So the butt of the ladder should rest about ten feet from the building. Or again a seventy-five-foot ladder should be placed approximately eighteen feet away, and so on. This rule is not worked out to inches, but it constitutes a fair guide when sidewalks are very wide or the reverse and affords the officer in charge a working basis.

A few words may perhaps be appropriate at this point anent an unique feature of American fire practice, namely the water tower. This apparatus consists merely of a lattice work tower mounted on a quadrant, through the interior of which passes a hollow tube, through which again an extension tube on the telescopic order is fitted in such a manner that it may be extended in a similar way to the extension ladder. When fully extended the tower is between 65 and 70 feet. When not working it is kept in a horizontal position, being raised when necessary either by hydraulic pressure or by springs. It is especially useful in fighting fires in high buildings and that its adoption is not general in Europe is due to the fact that local conditions are not as a rule such as to necessitate its employment. None the less the following details may be of assistance to fire chiefs in all parts of the world.

In many small fire departments, and for that matter in some of appreciable size, it is customary to use the bed ladders on trucks as water towers. Not being possessed of the latter apparatus, chiefs are obliged to use a makeshift in order to get a nozzle into an elevated position at fires where the ladder cannot be lowered against the building. A ladder pipe after the style of turret pipes used on wagons and water towers should be permanently attached to the under side of the ladder and securely fastened. A length of three-inch hose is attached to the pipe and strapped to the ladder with a siamese connection on the ground. Should a water tower be needed, this will effectively take its place. Run in two lines, connect to the siamese, raise the bed ladder to the desired position and the stream is controlled from the street by guys or a man may be sent aloft to direct it. If it appears cumbersome to have an entire length with the siamese attached, a short length of about ten feet may be used and it should be kept in position with the pipe near the end, and in case a water tower is needed the short length is connected with a length of three-inch hose carried for the purpose. This will bring the siamese to the street level and will give mobility to the line. In case there is no stationary pipe on the ladder, an open nozzle should be connected to a length of three-inch hose and laid on the ladder with the nozzle pointing through the rounds about two from the top. It must be lashed in position, and if the company contains a practical, or even

WATER TOWER FROZEN IN

an amateur, sailor he should be employed to make the knot. Landsmen ought to be well versed in making such knots in advance, as it is useless to instruct a man in such work at a moment when celerity is most necessary.

A word of caution should be inserted in connection with the use of aerial ladders as water towers. Although the writer is aware that a ladder so converted is still available, all ladders should not be used for this purpose. Persons in the building may be cut off by the flames or if the fire-fighters are pocketed a long ladder may be necessary in order to effect their rescue and judgment must be used in the number of ladders which may be transformed, although all that is necessary to do is to shut off the water, disconnect the lines, lower the ladder, ship the steering wheel and move the truck to the point where it is needed. But it must be remembered that this will take an appreciable length of time, and the fire, sweeping toward the imprisoned people, will not await the convenience of the department. Hence use caution before converting an escape into a water tower.

As to the success of the gasoline motor for hauling apparatus, there can be no question. The experimental stage has passed and that of certainty has arrived. The main point is whether the motor should be built in as an integral part of the apparatus, or whether it should be detachable and be in the form of a tractor. In some ways the latter seems preferable since, should anything happen to the tractor necessitating extensive repairs, it is easily disconnected, another tractor attached and the apparatus kept in service. It appears that this style of traction is coming into general use and the writer believes will be universally adopted both in fire departments and commercially, as time demonstrates its utility. It is beyond the scope of this work to enter into a detailed description of each part of motor apparatus, but rather to afford some general idea of what experience has taught.

A word may here be said anent the storage battery. Many fire chiefs express a strong preference for this style, and, like everything else, it is possessed of some good points and a number of bad ones. Thus the batteries deteriorate rapidly when standing still and are expensive to replace. For effective fire service it is necessary to have a set charging continuously, while for heavy grades it is not as certain as gasoline. On the other hand it is easier to operate and is certainly more reliable for starting away in response to an alarm. While still in the experimental stage, it is worth watching, for there is no saying what developments may occur which will remove its present disabilities and place it beyond the sphere of the problematical. The intention of the writer is not to express a decided preference for any particular type, but rather to indicate the strong and weak points of apparatus, leaving individual chiefs to assess the merits or otherwise of the machines dealt with, and thus form their own conclusions.

In operating water towers and extension ladders, that is in raising and lowering them, there are several devices, electrical, mechanical, spring and hydraulic, all of which have points in their favour. The electric motor can raise an extension ladder with rapidity and the spring is equally good, some even considering it more certain. Hydraulic power or springs can be employed in raising a water tower. As before emphasized, locality must govern choice and public money should not be lightly spent. In fact the best test is to consider whether, were the purchaser paying for the apparatus out of his own pocket, he would consider it wise to spend the money.

In the foregoing a brief description has been given of the major pieces of motor apparatus, and it is now advisable to consider their operation. Hose wagons and runabouts possess no technical features, with the exception that they should be strongly built and with sufficient power to carry the loads placed upon them over rough paving and heavy gradients. But the fact should be appreciated that though a piece of apparatus may be seldom in motion, when it is needed it must travel like the wind. All departments have recently shown a tendency to run the hose wagon, equipped with chemical tanks, ahead as a kind of scout. The writer thoroughly approves of this method. As has been frequently repeated, time is everything in fire-fighting, and a few seconds gained may prevent great loss of life and property. The fast moving wagon arrives on the scene before the heavier engine and attacks the fire with chemical lines, remembering in all cases to stretch two and a half inch lines from the nearest hydrants. On the arrival of the engine, •connection is quickly made to the hydrants and preparations commenced to start water in these lines immediately, should the condition of the fire warrant such an operation. The old adage "Never send a boy on a man's errand" should be borne in mind, and therefore never trust entirely to a chemical line. Although it may extinguish the fire and thereby decrease the fire loss, it is utterly useless should the outbreak assume great proportions, and under these conditions a good stream of water is an absolute requisite.

In residential districts where the houses are detached, chemicals will extinguish many fires with the least possible loss, but their use should not be attemped in commercial buildings where great quantities of goods are stored, until science has given us a far more effective gas than is at present available. In employing a chemical stream it should be directed low, as it is not the quantity of liquid which extinguishes the fire, but the gas arising from it which does the work. As it is evident that, in order to get the best effect from chemical streams, the gas should be confined, such streams are of no avail in the open. Where a stream of water is used, the correct method is to strike the ceiling with the stream on entering a room, which distributes the water. Should the entire contents of a small room be involved the water is spread out like a fan over the whole area, when, as the writer's old mentor used to say, "A dash will put it out."

The next points to be considered are engine and high

pressure streams. As the homeopathic and the allopathic doctors always disagree about the strength of a dose of medicine, so there is always a difference of opinion amongst fire chiefs as to the size of nozzles to be used and the pressure required in delivering streams at fires. In the old days the volunteers christened their machines "The Niagara," "The Cataract," "The Deluge," names to denote overwhelming power, as it was the idea to drown everything in sight. Even today there is evidence in some quarters that this desire still remains. Nozzles 2, 3 and even 4 inches in diameter are sometimes used! Naturally common sense should govern the matter, for though there are instances where one powerful stream may save the day, there are others where several small streams are far more effective.

Take as an illustration the case of the Equitable fire; the strong gale drove the fire toward Nassau and Cedar streets—the latter only 27 feet wide and the former about 45—and the flames raged along Cedar street for a distance of 375 feet and on Nassau street for nearly 200 feet. Here was a line of 575 feet to be protected in narrow thoroughfares where the high buildings had windows of plain glass. One, two, three or even four powerful streams would not have had the desired effect. It required a perfect water curtain along the entire front and the only way to accomplish this was to cover the buildings with a deluge from small streams, supported by two-inch streams from the water towers to drive the fire back. Such an effective curtain resulted from this method that not a single pane of glass was broken in the windows of the exposed buildings in the streets mentioned. It was as though a heavy musketry fire from an entrenched army, backed by a few pieces of artillery, had checked the advance of a storming party. That this was wonderfully successful may be gathered from the Underwriters' Report, which states, that it was effective to a remarkable degree, and the writer would add, was the only means under the conditions existing at this fire which would

SEARCH LIGHTS.

have accomplished the desired result.   This is the homeopathic method.

Now consider the allopathic side.   To give an example; last autumn an explosion took place in a sulphur works built on a dock in Williamsburg, Borough of Brooklyn. A strong southwest wind was blowing, and directly across the street were situated the oil yards of the Standard Oil Company, while at a distance of only 45 or 50 feet were extensive hay sheds, one thousand feet in length.   A more dangerous combination is difficult to conceive!   A fourth alarm was immediately transmitted by the District Chief on his arrival, followed by a borough call which brought another third alarm assignment.   On the writer's arrival he found a most dangerous fire confronting him.

The entire building of the sulphur works was involved, the flames shooting a hundred feet into the air.   The hay sheds had ignited and the flames were rolling over the oil tanks to leeward.   Deputy Chief Langford, with eight engine companies and a fire boat, was assigned to the leeward position in the oil yards and six companies under Deputy Chief Lally were placed in the street between the hay sheds and the latter.

Although the water supply was ample, the combined force of these fourteen companies using powerful streams was not sufficient to drive back the wave of fire which momentarily threatened to envelop the oil yards and bring destruction to life and property.

The fire boat, Abram S. Hewitt, had worked her way in to within 75 feet from the head of that threatening and destructive sea of fire.   She was operating about eight $1\frac{3}{4}$-inch and $1\frac{1}{2}$-inch streams.   The order was given that all streams should be shut down with the exception of two $1\frac{1}{2}$-inch, which were operating under Langford in the yards, and that a three-inch nozzle should be put on the large monitor on the top of the pilot house.   The full force of the pumps was then thrown into the nozzle, giving a pressure of 145 lbs. and a discharge of more than 3,000 gallons per minute.   The effect of this was to crush the head of that fiery wave and roll it back and hold it, giving the men operating the smaller streams a chance to advance. Compare the two examples. In one a threatening fire had spread over a great area; in the other a terrific wave of flame was concentrated within a narrow space. A cyclone in the first instance; a tornado in the second.

So it will be seen that in fire-fighting both homeopathic and allopathic methods are required, but if the fire can be extinguished with a small stream a large one should never be employed. How best to judge the size of the dose the writer is unable to tell. As each doctor diagnoses his own case, so each Chief must make his own working diagram, and, like some physicians, many fire-fighters have better discernment and keener judgment than others, sometimes born of greater experience, and sometimes more or less intuitive. All countries are looking for good men and paying liberal salaries, and each must study and fit himself for the ordeal. No maps or charts are available, the surroundings must be noted and decision must be prompt and effective.

The intelligent operation of apparatus by members of fire-fighting forces is an absolute essential successfully to cope with their enemy. The direction of streams at a fire is almost akin to gun fire from a battleship. Shot and shell can be, and indeed very often are, wasted, due to defective gunnery; streams thrown into a building at an improper angle are useless when, by a little judgment on the part of the officer in charge, much more effective work could be accomplished. This is particularly true of water towers and turret pipes. In placing a tower in front of a building, a little quick thinking on the part of the officer in command of the apparatus would often make a great difference. To begin with, a tower should never be placed far to windward except under orders from a superior. Ninety-nine times in a hundred the fire will work to leeward, so the tower should be in a position whence the fire can be fought back from that side. The mast should be extended high enough to give the stream an arc which will enable it to strike the ceiling of a particular floor about 20 to 30 feet inside the window. As in gunnery practice the best arc is about 40 degrees elevation, but as a tower stream is not directed at a target but must be operated over the entire front of a building, judgment should be used in its placing and elevation in the beginning.

In connecting lines to a tower, deck pipe or turret pipe on a wagon, about one length—50 feet—of extra hose should be allowed. This is to give the apparatus a certain degree of mobility, as it often becomes necessary to move a tower or a wagon, and should there be no spare line it cannot be accomplished. Officers ordered to connect to water towers or wagons should bear this in mind. Too much rigidity is undesirable, and by having 50 feet of range in front of burning building, much more effective work can be accomplished especially in these days when motor apparatus is so easy to move.

While on this theme it may be wise to touch briefly on the size and pressure of streams. The use of a water tower, or of a wagon turret pipe, presupposes the necessity for water under high pressure. Some chiefs have contended that a higher pressure than 100 lbs. at the nozzle is impracticable, saying that when the pressure goes over this nozzle velocity it is so high that the stream is torn to pieces (whipped into a foam) causing it to break and scatter a few feet from the nozzle. This contention is only partly true, for, admitting that the higher the nozzle velocity the greater the tendency of the stream to disintegrate, on the other hand the higher nozzle velocity gives a greater volume and a much heavier striking force. These latter are prime requisites in fire-fighting. A blow struck or a thrust delivered with moderate strength has not the same effect as a blow with good muscle behind it, and in big fires it is often the power of the blow which counts. The stream which strikes with force will knock out the fire as a prize fighter knocks out his opponent. The writer favors higher pressure when necessity demands, though this is again a matter of judgment.

In the army and navy, range finders are employed to enable officers in charge of guns to sight properly and direct their fire. Fire-fighters operate under such conditions that many times they are unable to see even the buildings, much less the windows or openings through which the streams are to be directed. In such cases all must be left to judgment and experience. When a heavy body of fire is rolling out of a building the writer would use a nozzle pressure of from 100 to 150 pounds. This gives volume and that striking force, which, as already stated, are absolutely necessary in many instances.

Before passing from the subject it may be well to give a rule by which men may be guided in placing apparatus of the type under consideration. Good firemen fight at close range, and it is seldom that the occasion arises which keeps them at a distance, but under such circumstances a stream will do effective work allowing one foot of distance for each pound of nozzle pressure up to 100 pounds. Above this point it is a little less, but it is useless to go further into the matter as the occasion should never arise when it is necessary to fight a fire at a greater distance than that of 100 feet.

In some instances the actual distance covered by the stream is greater when the nozzle is at, on or near grade level. The point of delivery may be 50 feet above grade level, and those who care to amuse themselves and at the same time improve their knowledge, can use the old rule; the square root of the sum of the squares on the two sides of a right angled triangle is equal to that of the hypotenuse. The main point is to know the distance to which a certain pressure will deliver effective fire streams. Hydraulic engineers like J. F. Freeman and others give distances, always under settled weather conditions, but a fire chief cannot wait for a calm day, so the confronting conditions must be figured out. A little thought will make a most wonderful improvement and it will be found

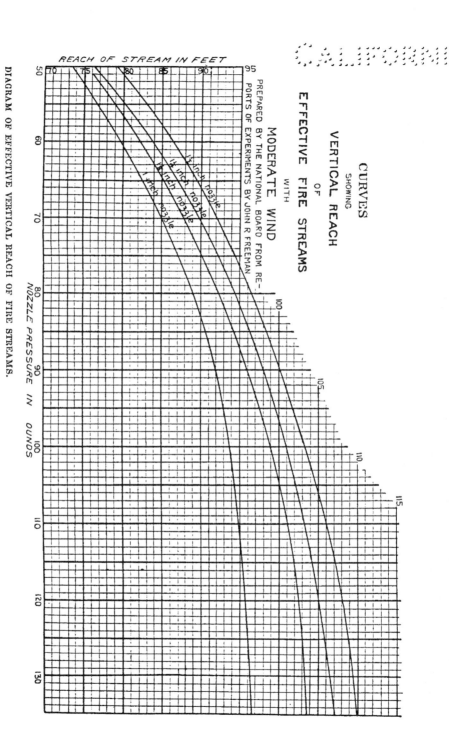

CURVES
SHOWING
VERTICAL REACH
OF
EFFECTIVE FIRE STREAMS
WITH
MODERATE WIND

PREPARED BY THE NATIONAL BOARD FROM RE-
PORTS OF EXPERIMENTS BY JOHN R FREEMAN

REACH OF STREAM IN FEET

NOZZLE PRESSURE IN OUNDS

DIAGRAM OF EFFECTIVE VERTICAL REACH OF FIRE STREAMS.

by experience that men can estimate within a few pounds of the exact pressure at a nozzle. This will be treated exhaustively in the appendix.

In dealing with the effect of additional lines in the tower, deck or turret pipes it must be stated that, in most departments, the orders require a company stretching to the tower, deck or turret pipes to lay in two lines. Now, if a two-inch nozzle is used on the tower mast, much better streams will be obtained by immediately adding another three-inch line. It may be assumed that a nozzle pressure of 100 lbs. is required from a two-inch nozzle; this is equivalent to a discharge of 1,200 gallons of water per minute. In order to obtain this pressure, each of the first two lines would be delivering 600 gallons per minute, and the addition of the third line would cause this flow to be reduced to 400 gallons for each one. As the friction loss is equal to the square of the velocity, and as the velocity is governed by the flow, the result would be something like this: $400 \times 400 = 160,000$. $600 \times 600 = 360,000$, or as 16 to 36. Therefore friction loss would be more than twice as great as in the first case.

This takes the writer back to the source of supply, whether it be a fire engine, a high pressure pump or water under force of gravity which becomes the determining factor as to what that pressure should be. If it is decided to force 1,200 gallons per minute through two lines, the pressure must be higher at the pump than if the same quantity were forced through three lines, in order to give the same nozzle pressure. In the case of water delivered under high pressure by pump or force of gravity, where there is a predetermined pressure, the additional line gives a better flow, less friction loss and consequently a higher nozzle velocity. The chief officer must determine whether it is better to attach the additional line, thereby taking up the services of an engine, or to use the third opening in the hydrant to which a line might be connected for service at some other point, where it was badly needed.

In this chapter the writer has had no thought of actually

discussing fire strategy; all he has hoped to do has been to point out some of the situations which develop at fires, and to impress upon those responsible that their minds should act quickly and that **all** contingencies should be met with promptitude.

CHAPTER XXIII

•

ANY consideration of the factors governing fire-fighting would be incomplete were the personal equation neglected, in other words the status of the fireman in relation to the community as a whole. The foregoing pages have made it abundantly evident that he who wishes to follow the calling as a life's work must be possessed of qualities out of the common. Apart from pluck and physical endurance, he must be prepared to attack the subject from its scientific standpoint and to devote to it his undivided interest and attention, that is to say if he would rise superior to his volunteer predecessors. And it must never be forgotten that these latter brought unlimited enthusiasm to their self-imposed task, if nothing else. But in these days, fire-fighting has passed the stage of the dilettante; it has grown into a serious science and as such has a right to demand that its votaries shall be experts in their own line and that the huge responsibilities of lives and property entrusted to their care shall be no haphazard proceeding, but the result of serious consideration and selection. Now latterly a feeling has sprung up, which roughly propounds the theory that the fireman should be the hired servant of the municipality, nothing more and nothing less. His engagement is to be of a temporary nature, bounded only by so many hours' work *per diem* for so much pay. He is to be considered, when not actually on duty, as free as the butcher, the baker and the candlestick-maker and from the moment when he leaves his fire station, what he does and how he does it is to be his own business and no one else's. This is to all intents and purposes what the advocates of the "Two Platoon System" would have the public believe is to the advantage of the Fire Departments the world over. For twelve hours the fireman is to be on duty, and for the other twelve he is to be a private citizen, free from all trammels and restrictions, which might tend to hamper his sacred liberty. It is a high sounding and truly attractive doctrine, only, having due regard for the efficiency of this most important of departments, how can such a method be expected to yield satisfactory results? In France and Germany, the barrack system is worked no doubt to an extreme, which would be distasteful and unpopular in any country where conscription was not the general practice. There, the fireman is practically under military law, and the slightest breaches of discipline are punished with martial severity, whilst individual freedom is necessarily much curtailed. But there is no advantage in riding a hobby to death and both in England and in some cities in the United States, a *via media* has been evolved, different in detail but alike in essentials, which insures efficiency of service and discipline with the least possible inconvenience to those concerned. In England, the fireman is a municipal servant, he is provided with quarters whether married or single and is granted so much leave every year. In addition he receives free medical attention and certain allowances according to the town in which he happens to be serving. But all the time, when not absent with permission, he is available for duty, as is the soldier or the sailor. This appears a common-sense system of enabling a municipality to assess with some degree of certainty the probable fire risks it will be called upon to incur, should it either increase or decrease the fire department, while it can always count upon the services of its enlisted men. New York has gone further and allows its firemen to live at home, have their meals there and to spend their spare time in the society of their families, far removed from the noise and bustle of the average fire-station. In fact, the system evolved may be taken as one of encouragement to home life and happy surroundings, the antithesis of arbitrary militarism and the curtailing of the individual's freedom. Only the demands of the situation are such, that it is imperative that at all times and seasons, the men of the department, except when specially excused, should be available for instant duty. Now though comparisons may be odious, what alternative system is advocated by the upholders of the two platoon theory? For twelve hours one watch or division of men is to be employed, their places to be taken by another group of similar strength. And supposing some great conflagration occurs, demanding the combined energies of the entire department, what then? Presumably messengers would have to be employed to whip up the contingent off duty, who from their standing would be within their rights in demanding "overtime," who perhaps could not be found and if found might demur to answering a call which was not obligatory, and who from their anomalous position might be forgiven if they imported an opera bouffe touch into subsequent proceedings and kept a fire burning so long as might suit the financial status of those principally concerned. In addition what guarantee would be forthcoming that the night shift would arrive in the best of conditions to carry out arduous work, if called upon so to do. It has been proved from experiments made along these lines that it is by no means uncommon for the so-called fireman to practise his trade as bartender or undertaker all day and then turn up at the station in the evening for fire duty, in order to make a little "extra." It goes without saying that that species of extra time is costly alike to human life and property and that a fire department built upon such principles is indeed founded upon sand. The truth of the whole matter would appear to be, in the opinion of the writer, that amongst the younger generation there is a strong aversion to discipline, even of the broadest kind, which interferes not at all with the liberty of the subject and is only ordained that some systematization of duty and responsibility may result. Further, it is of paramount importance in these days of labour unrest, that any body so vitally essential to the welfare of the public at large should be safeguarded from any interference from labour organization outside itself. Once the Department becomes analogous with street railways or drygoods stores for its labour supply, then a long farewell may be said to any continuity of action, of policy, of training, and of general organization. Men will come and go at will, bearing in mind that their status is no different from that of any other daily worker, *esprit de corps*—a vital force in fire-fighting—will vanish, and in place will be instituted a body, devoid of any other consideration than that of making that little bit "extra." An inspiring ideal, in truth.

RESUSCITATING FIREMEN OVERCOME BY SMOKE.

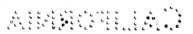

CHAPTER XXIV

UNDERWRITERS AND SALVAGE CORPS

ALTHOUGH not actually connected by organization with the Fire Department, at least in the United States and England, the Salvage Corps is none the less an important essential in the whole scheme, and as such demands some attention in a volume of this kind. The business of the fire-fighter is, as his name indicates, to fight fires; the saving of property and merchandise from the effects of water is the particular field of the Salvage Corps, and thus it is that separate organizations have been evolved for this express purpose.

In New York this work is officially recognized and supported by all the insurance companies doing business in the city, supreme executive control at fires being vested in the officer in charge of the fire-fighters. In fact, the Salvage Corps might not inaptly be compared with the medical and nursing branch of an army. A battle occurs, and obviously the combatants have no time to look after the wounded or to succour the dying. Their business is first and foremost the crumpling up and the defeat of their antagonist, and until that is accomplished all other considerations are relegated to the background. But humanity demands that the stricken shall be saved if possible and that suffering shall be alleviated; hence the presence of a large staff of doctors and nurses who otherwise have no connection with the events of the moment. They are distinct from the fighters, but admittedly necessary to them. Further, of course, they are under the supreme authority of the general commanding; anything in the nature of divided control would be fatal alike to all concerned.

This brief explanation is rendered necessary since the functions and position of those operating in salvage corps is too little understood, and in fact their present status of a semi-independent entity is anomalous and not in the best interests of all parties. Now who are the persons involved? Firstly, property owners themselves who, whether insured or no, are at least presumed to desire as little destruction wrought to their goods as is possible, which can often be accomplished by covering up, wiping down and pumping out. Secondly, the insurance companies, since, with this assistance, they may save the wastage of a complete loss and at any rate realize a small proportion of the risks involved. Hence it is that a regularly paid corps of professional salvage men is maintained at the expense of these companies in some of the great towns of America and in London, those in the United States being generally known as "Fire Patrols."

The strength of the New York corps consists of 38 officers and 190 men, divided into ten patrols. The salvage wagons in service are of the usual type carrying such apparatus as covers, door openers, axes, shovels, squeegees, and other equipment especially suitable to their needs. During the last year for which figures are available 8,415 outbreaks were attended, the most calls in any one month being 802 in December. In that month the duration or service amounted to 755 hours, while 3,531 covers were spread and 136 roof covers placed in position. Chicago maintains a corps of eight companies consisting of 71 permanent, and 33 auxiliary, men, the work accomplished, having due regard to the territory involved, being efficient.

Incidentally, annual reports are issued giving statistics anent the forces together with figures of the values they have protected, which, of course, is as it should be, and is only mentioned as a comparison with the extraordinary anachronistic condition of affairs in London. Here is the largest city in the world with property values almost beyond estimate. It is asserted that the "per capita" loss is insignificant compared with that of American cities, but at the same time it would be imagined that salvage service would be carried out upon certain well established lines. Yet, as a matter of fact, the London Salvage Corps, to give it its official title, is a purely private organization, equipped and supported by certain of those insurance companies underwriting fire risks. The word "certain" is used advisedly, since not all those dealing in the local fire risks subscribe. Hence it is a purely private undertaking, and, as such, responsible to no public body, except in the same degree as a private individual. No annual report is published, except to those directly concerned, the strength of the corps in officers and men is not officially issued, and any statistics of property values, percentage of losses, etc., if tabulated at all, are certainly not forthcoming from this body, at present benignly commanded by Colonel Fox. Of the annual report, such as it is, the following excerpt from a letter speaks for itself: " . . . at the end of the year a brief confidential statement is given out to the members, which statement is not illuminating in details." In other large English cities, salvage arrangements are altogether lacking, as is also the case in the smaller American centres of population. The reason for this is not far to seek. If the insurance companies have to pay the piper they cannot be expected to establish in communities with minor property values, organizations which, on the face of it, must be costly experiments.

Hence, two clauses of objection may be framed upon the existing state of affairs, which cast no slur upon those intimately concerned but are simply concomitants of an archaic system. Firstly, no matter how cordial the relationship existing between the fighting force and the salvage corps, and no matter how rigorously it is insisted that the Chief is in supreme control at an outbreak, with the best intentions in the world, on occasion, misunderstandings are bound to arise which, if not hazardous, are at least unpleasant, unnecessary and not in the best interests of discipline. There is always the natural tendency for the subordinate to resent taking orders from any other than his superior officer, and though the Chief and his aides may be recognized by the non-combatant forces as in executive command, at a large fire neither the chief nor his staff can be everywhere, and orders have to be transmitted which, to comply with the hard and fast etiquette of the situation, would be cumbersome and impossible. Therefore it would appear a common sense policy to arrange some form of amalgamation. This should offer no insuperable difficulty, as some method of meeting the additional cost could be evolved in consultation with the insurance companies.

Which introduces the second clause.

As at present arranged the maintenance constitutes a severe tax upon those concerned, and there is no doubt that better administration would result from the union. In addition, there is no reason why the individual property owner should not bear his share of the costs involved in the saving of his goods from unnecessary damage by water, consequent upon fighting a fire. Of course it may be argued that if he is already insured, he is paying twice, but the adjustment of premiums under such circumstances would be a matter of no great moment.

Certainly, in New York and in America generally, there is this to be said for the salvage corps, namely, that they are responsible bodies maintained by statute and thus under the supervision of recognized authorities. In England, however, there appears to be a total absence of organized departmental control, and the relationship existing between the London County Council, which supports the London Fire Brigade, and the London Salvage Corps, which is as much a private concern as William Whitely or John Wanamaker, is purely dependent upon the good will of the former. In fact, to go a step further it is hard to see what means could be adopted if the County Council vetoed the presence of the salvage corps at fires. Of course such an occasion would never arise, only, in marshaling fire forces, as in marshaling the array of an army, nothing should be left to chance, nothing should be haphazard and all contingencies should be foreguarded. And since the general commands a fighting machine composed of combatants and civil units, the admiral a personnel, a proportion of whom are either clerks or semi-laymen, so should a fire-fighter and his non-combatant ally serve under one flag, recognize one chief and be brothers in a common cause. There is no good reason why this should not be accomplished, and that till now it has not been consummated has been due largely to the apathy of those in responsible control. Any general consideration of the problems connected with fire insurance is quite beyond the scope of this volume. At the same time, however, it may be of interest to quote certain comparative data and, in addition, the writer is happy to give prominence to the good work being accomplished by underwriters in the United States.

It is a common failure in America, generally to condemn certain social factors as being responsible for this, that or the other occurrence. The acute observer will have noticed that under no circumstances is the individual unit ever responsible for the sequence of events; come murder, rapine, pillage or earthquake it is always that inconsiderate "somebody else" corporate or incorporate, who is to blame. Never under any circumstances, has that wonderful individual unit been in error. And the same might appositely be said of the press. Clever writers with brilliant pens overflowing with ink, between eleven o'clock at night and two in the morning, will, with the utmost complaisance label corporations or persons in a genial way as "rogues," "vagabonds" or "villains," excepting always that their readers will take their remarks with that proverbial grain of salt which is necessary to a true appreciation of the journalistic cuisine. Unfortunately, the average reader is stolid, unsuspicious and indiscriminating. He rarely sees beyond his nose, he never reads between the lines and he

is as a child in the complicated moves in the game of
publicity and policy pursued by many. Which remarks
may be accepted for what they are worth. Few acquainted
with public life will honestly contravert them, and for the
rest, their opinion is of equal value too, though of far less
weight than that of the writers mentioned. Thus it is,
that a compliment must be paid to the National Board of
Fire Underwriters, which, in the United States, is doing
a great deal toward the education of the general public
in the matter of fire risks.

The following excerpt taken from the address of an
officer of the Northern Assurance Company of London,
is worthy of quotation: "The phenomenal growth of
this country (the United States), of which all its citizens
are proud, has been accompanied by the creation of enormous commercial and manufacturing establishments, great
concentration of property values, and congested areas in
our large cities, each containing hundreds of millions of
dollars of values subject to destruction by fire. This extraordinary commercial and manufacturing growth makes
necessary a corresponding increase in credits. To meet
this need for larger credits the banks have been made larger
and are being made larger all the time. One of the most
important and essential foundations of credit is fire insurance. This is now so well understood that it is unnecessary to dilate upon it. As the great growth in business
requires a corresponding increase in credits which in turn
makes necessary larger banking facilities, so the concentration of values all over the country in individual businesses
and in congested city centres makes it necessary that fire
insurance companies should have large and increasing
reserves and loss paying abilities. Under a general system
of State-made rates, fire insurance companies would undoubtedly find themselves unable to build up large reserves
and there would be no inducement to make them large by
capitalization. The welfare of the country requires that
fire insurance companies should have the opportunity to
create large reserves with which to meet large conflagration

losses such as have several times occurred and must be expected to occur again. Inability to meet such crises would cause serious and perhaps dangerous panics. The average underwriting profit made by all the companies has been paltry, as will hereafter appear, and furnishes no justification for State Rating Board laws on the ground of excessive profits. The smallness of the average underwriting profit shows that the dividends to stockholders have been paid from interest and dividends from investments, and also that reserves have been augmented from the same sources. It is, therefore, apparent that excessive rates have not been charged and the proceeds distributed to stockholders." The above may be an "ex-parte" statement, but on the face of it, it contains a thorough realization of the condition of the fire risks in America.

Time and again it is insisted by the underwriters that it is the conflagration hazard which renders any approach to systematic or scientific underwriting practically impossible. If, as in Vienna, it were feasible to confine every fire to the building in which it originated, which can obviously be accomplished when houses are constructed upon a detached system, then the problem would be simplified. If, as in Germany, a man having a fire on his premises is held guilty of misdemeanor until he proves that its inception was beyond his control, or if, as is operative in France, a man having a fire extending beyond his own premises were held financially responsible, then again the problem would be simplified. But, probably, the sanest and most hopeful method of combating the huge fire losses in the. United States is by the gradual instruction of the masses to a proper realization of fire control and what it means. To this end the National Board of Fire Underwriters has most distinctly added its quota of assistance by the publication of useful brochures and by the issuance of bulletins apropos to certain occasions when fires most commonly occur such as Christmas Day with its Christmas trees and candles, and Independence Day (July 4th) with its indiscriminate use of fireworks. No one, however partizan, can deny that this is useful work and of the greatest public benefit. Annually the fire loss in the United States may be roughly assessed at $250,000,000 or 59,999,999 pounds sterling. This is naturally only a broad estimate, for to be irritatingly accurate the estimated loss for 1911 was $217,000,000, though in 1906 it was over $500,000,000, the latter figure, to be sure, due to the San Francisco disaster. But taking even $200,000,000 as the annual wastage, it would be profoundly interesting to know how much of that colossal total was due to the carelessness of the individual. It is beyond the ken of man how to arrive at such figures, but the writer can testify from personal experience, and has emphasized again and again in the preceding pages, that the human element plays a preponderating part in providing work for the fire-fighter.

There is always something fascinating in picturing in the mind's eye all the wonderful things which might be, were such and such a factor eliminated from the social fabric. The pacifist who dreams of disarmament and international peace is wont to enlarge upon what the reduction of the Army and Navy estimates would mean to the masses. A secretary of the Treasury, or Chancellor of the Exchequer with a hundred million dollars unallocated to any particular purpose in these dreary days is as rare as the "Dodo," in fact he is non-existent, but supposing some such amazing *bouleversement* of conditions did occur. It is only possible to surmise how that money might be spent for the good of the community. It appears on the face of it so fantastic, so irrelevant to the issues of the moment and so far removed from practical politics. And yet implant in the individual the fundamental features of the dangers of fire and of the easiness of those ordinary precautions which in themselves are simple to the verge of puerility, but in the aggregate count for so much, then there might be a reasonable chance of the materialization of that seemingly far-fetched dream.

Hence, all means to that end are to be encouraged, **and**

the work being accomplished by the National Board of Underwriters is deserving of more than passing praise on that score. For, that there is need of some guidance, can be gleaned from the following figures, which, making every allowance for climatic extremes, as shown, a fruitful source of fires, and constructive encouragement to the flames in the shape of the employment of wood as building material, still evidence a remarkable gulf between fire risks in Europe and the United States. Thus, whereas the "per capita" loss in the United States for the year 1911 was $2.62, for England it averaged only 53c, for France 81c, for Germany 21c, and for Russia only 8c. These figures are not absolutely conclusive, since they are based upon the reports sent in from centres of population and no account can well be taken of the country in sparsely populated districts. But as regards the United States, it may be remarked that the 298 cities upon which the average is framed, represent a population of 31,000,000 people, among whom there is "fire protection." In no foreign city does the "per capita" loss approach five dollars (one pound), though, excluding conflagrations, 32 cities in the United States surpass that figure. The highest fire loss in England, curiously enough, was in the ancient city of York, which, with a population of only 82,000 shows a "per capita" loss of $2.73 (11/—), which compares unfavorably with the town of Yonkers, near New York, with the same population and a "per capita" loss of $1.73. For London the "per capita" loss was 54c based upon a population of seven and a quarter million people, which evidently included its suburbs and not the county proper. The "per capita" loss in Paris was 60c, for Hamburg only 18c, for St. Petersburg 93c, for Moscow $1.46 —exemplifying excellently the rise in loss occasioned by extremes of climate and wooden buildings—while the city of Vancouver heads the list with $2.61.

These are statistics which cannot fail to provide food for the thoughtful. For fire wastage is a literal translation of the phrase commonly used to emphasize the extravagance of the individual, "burning money." Finding money for the provision of armaments, which may never be used, is often decried as hideous waste, though there is something tangible to show for the expenditure, what then can be said in defence of a loss to every man, woman and child in the United States of over two dollars per annum, with a visible result only of charred ruins and possible suffering?

CHAPTER XXV

CONCLUSION

It is the experience of most writers on themes peculiarly their own, that when the time comes to surrender the pen, only then do they realize how much more should have been added to the material collected, how much more clearly certain details might have been explained and how altogether inadequate is the sum total of their labours. And so it is as regards this work. The term "fire-fighting" possesses such an immense significance that it would require many portly tomes to deal exhaustively with all its intricate problems. For as fast as one offshoot receives attention, another crops up hydra-like and it becomes a question of serious consideration what proportion of notice, if any, should be allotted to many subsidiary branches of the main theme. Thus it is only feasible by using judgment to scrape the surface of the vast field of investigation, and, by as skilful handling as may be to sow the seeds of fresh thought in the public mind, which in time may germinate and bring forth a rich harvest of action. This indeed has been the chief inspiration of this volume. Those interested scientifically will, it is hoped, have found the wherewithal to whet their appetite for further research and there is no doubt that there are problems without number awaiting the probe of the investigator in the domain of "fire control," problems which demand the maximum of mental efficiency, patience and determination. But the larger portion of the reading public demand the note of human interest and as far as lay within the scope of the material, this has been supplied, for it is a trite commonplace, that

347

to arouse enthusiasm or a proper realization of the true inwardness of any subject, appeal must be made to the senses.   And it appears of paramount importance to the writer that "the man in the street" should be aroused from the lethargy he habitually displays over questions appertaining to fires and their prevention.   Terrible lessons are taught, and as speedily forgotten, not because the individual is unable to grasp the significance of the occurrences, but rather because the teachers are intangible elemental forces which cause disaster and await the skill of the ·human instructor to amplify and explain their lessons.   And it may be said that until recently such instructors were sadly lacking.

In the first place the search after wealth is so keen now-a-days and monopolizes attention to such an extent, that occupations promising only a moderate financial sufficiency are to a certain degree shunned by the enterprising and the ambitious with the result that those enlisting in their ranks are often devoid of that enthusiasm, the magnetism of which is so overwhelming that it communicates itself to others and sweeps into its net the youth and strength of a nation.   With such material, what can be expected? Any individual adopting a calling merely as a means to obtaining the bare wherewithal to exist rarely earns his wage.   His duties are rendered just so efficiently as to escape censure and no more.   Now, without any savor of supreme conceit, the writer hazards the statement that for a change towards better things the fire departments in the United States are responsible.   In the first place the magnitude of the risks involved has impressed itself upon the thinking portion of the community and it has at length been realized that if an army and navy are necessary adjuncts to the safety of the Republic, then most assuredly the organization of a first-class fire-fighting force is necessary for the safety of the homes of the people.   And that is the chief factor in the situation; financial losses may be huge, but they will not appeal to the heart of a community to the extent of the death of one child by the flames.

To deal seriatim with the crucial points of each chapter in the foregoing pages is out of place in this conclusion, since if each in turn has not been sufficiently conclusive then the carpenter in this instance can not blame his tools, but himself. However, it may have been noticed that on occasion, comparisons, those most odious of things, have been instituted, while criticism has seldom been far distant when dealing with foreign departments. It almost seems superfluous to state that whatever has been penned has been actuated by no carping spirit, but rather to bring out some obscure point or to emphasize some necessary moral. If every fire department was organized on precisely similar lines, then assuredly the fire fiend would play more havoc than it habitually does, since in fire-fighting as in other spheres of life, one man's meat is another's poison, and most assuredly what will suit the requirements of one locality will most hopelessly fail in another. Further, criticism is a life-giving sap to any industry or profession, or, for that matter, to any human enterprise. Devoid of this spur, achievement would quickly cease from sheer inanition, and inasmuch as trees from time to time need pruning for their better growth, he would be a sorry gardener who hesitated over the task, lest perchance he might injure some branch or offshoot. Hence the writer hopes that all his remarks will be accepted with the same good nature as that with which they are offered. For there is need today for cohesion and mutual support amongst those, who have made, and are making, of fire-fighting, their life's work. This applies to all countries without exception, though there may be some slight difference in usage and organization. It is a peculiar reflection on human character and not altogether a pleasant one, that recognition of services rendered is too often overlooked, unless there is an accompaniment of glittering uniforms and the blare of many bands.

Stand in the limelight, make use of every channel whereby publicity may be gained and *ipso facto* the end may be attained! From mouth to mouth spreads this spurious fame; municipalities recognize it and governments decorate for it. Now those in the fire-fighting profession, though discounting the hollow adulation of the ignorant and assessing at its proper worth an evanescent popularity which is of no permanent value, may be forgiven if at moments they yearn for some sincere acknowledgment from those whose lives and property they day by day protect. Of course, it may be argued by some that they are paid for their services, which is all they should expect, but there are equally those who will query if their wages are equivalent to a daily risk of life and limb, unknown even to the professional fighter, the soldier and the sailor. As a matter of history, probably every fireman in every fire brigade in the world is only too ready and too willing to go to any extremity in order to save a human life. Surely it is not pleading for too much, if then there may be some sympathetic recognition of the fact, without the aid of brass bands and cheap publicity.

Of course, the press agent, that questionable offshoot of modern journalism, will accomplish wonders. And be it added, that, though not blessed too liberally with this world's goods, there are those who regard their dignity as above such methods of currying renown. This is particularly written anent fire-fighters of every rank and nationality. They do their work and if a grateful community is too busy or too engrossed with its own affairs to mind their deeds, their sacrifices or their welfare, then they accept the situation, though they may be forgiven if at moments it does seem to them a little callous, a little unfair and not altogether considerate. Reading an English newspaper the other day the writer came across a story of simple heroism, which he may be pardoned for reproducing with some detail. It is by no means an uncommon occurrence, it possesses no limelight effect and the surroundings of the event by no possible stretch of the imagination could be described as exhilarating or of a nature to inspire desperate courage. Rather are all the facts the reverse of these, but the narrative exemplifies the silent heroism which is no monopoly

NEW YORK FIRE STATION. NOTE SLIDING POLE, AUTOMATIC HARNESS AND ALARM BOARD.

of any fire department in any country, but is rather the heritage of precedent handed down from time immemorial and which carries with it that glorious superscription, "for a greater thing can no man do than that he lay down his life for another." The excerpt is given practically word for word. "That heroism is by no means confined to the battlefield or to the wild corners of the world's surface was exemplified once again by what happened in London on Tuesday, when the gallantry and cheerful self-sacrifice of the men of the London Fire Brigade was displayed. Two of them lost their lives in a splendid attempt to rescue a number of workmen overtaken in a rush of poisonous sewer gas under the road in Pembridge Street, Notting Hill. The men were employees of the Kensington Corporation. One of them was suffocated before he could be reached and four of his comrades came within an ace of suffering the same terrible fate. They were all struck down unconscious before they could be hauled out. . . . A dangerous and threatening leakage of gas was the cause of the trouble. This was reported to the Kensington authorities some time before noon and five sewer men were immediately despatched to locate it. Dressed in their heavy uniform and armed with safety lamps, they descended the man-hole in Pembridge Street to search for the source of the trouble, leaving as is customary a mate on sentry above ground. They had not been down long before one of the men reappeared at the top of the man-hole livid and gasping for breath. He was only able to gasp out that his mates were "knocked over" by the suffocating fumes before he collapsed on the pavement. A message was immediately sent to the fire brigade station and a large crowd assembled as the firemen dashed up from Bayswater, Notting Hill, Kensington, Euston and other stations. There were scenes of great excitement as the men, with their smoke helmets adjusted, disappeared one after the other into the man-hole. Amid rousing cheers three men were brought up livid, gasping and only just alive. It was known that one man, Parry, was still somewhere down there in that horrible atmosphere and the most desperate efforts were made to get to him. Fireman after fireman went down only to be dragged back exhausted after a futile effort to reach the entombed victim. They must have been pretty sure that by this time he was dead, but that made no difference. Dead or alive, it was their duty to get this man out and they stuck to their task with splendid heroism. The fatal tunnel was very narrow, at no part higher than three feet nine inches. Parry was a man of considerable bulk and it was thought that he had become wedged. The poison down below was worse and far more penetrating than the smoke of an ordinary fire, but the firemen cheerfully took the risk. All the afternoon the gruesome search went on and at half past four the anxious watchers at the top were signalled to haul up one of the firemen. He was dragged out nerveless and dreadfully pallid, and though he was treated on the spot by Dr. Kennedy it was too late. Half an hour later another was hauled up the narrow iron ladder and laid gently on a tarpaulin. He was found to be past all human aid, and the lookers-on bared their heads as the body was placed on a motor engine and taken to the mortuary. Traffic was diverted for many hours, the work of rescue being followed with breathless interest by a silent awe-struck crowd." It might be well added that the names of these two self-sacrificing heroes were William McClaren and R. Bibby. Now the object in having given these details in extenso is to emphasize as well as the ability of the writer will allow, the daily risks of the firefighter, often, as in this case, devoid of that supreme excitement which prompts many a deed of "derring do" upon the battlefields. And for those who accomplish the latter there are the thanks of a grateful country, or, at least, that is the official phraseology. At all events there is recognition of sorts not the least mark being the esteem in which those comprising the fighting forces are naturally held.

There is nothing to be gained by further following the line of thought. It has ever been so and unless some vast change sweeps over nations and peoples, it will ever remain so. The student in his closet, the chemist in his laboratory, the surgeon in his consulting room and the engineer in his workshop, all may devise some great benefit for humanity, but if they expect earthly reward they most assuredly are mistaken. They must look elsewhere for that, for they have no trumpet blowers and the intelligence of the individual as a rule does not stray beyond the pages of his favorite paper, which formulates his ideas, prearranges his opinions and supplies him with arguments along almost automatic lines. Hence it is as well that all connected with the fire-fighting profession should leave notoriety to others to whom it is as the breath of life and who find solace in well turned laudatory phrases rather than in a quiet conviction that they have done their best. The student of history must sometimes wonder how men of the stamp of Lincoln, Washington, Benbow, Burke, Wellington, and of course that immortal hero, Nelson, would have thought of the brass brand, press agent, limelight form of publicity, which today is so often regarded as the hall mark of meritorious service. However, the mole, if it works unseen, accomplishes a deal of tunnelling as is eventually ascertained, and the fire corps of the world though the sphere of their usefulness may not be readily appreciated, can afford to await developments.

There is, however, springing up in many quarters a proper realization of what fire risks actually are, and to that end private organizations are being instituted for the scientific and careful study of the problems involved. This in itself is a healthy sign, while annual fire conventions, of which one is to be held in New York this year, do a great deal towards advancing the interests of the fire-fighter and arousing the attention of the public in his career. And, in addition, since a rope is made of many strands, these conventions assist towards international friendship and a better comprehension of national characteristics. There is a talk of international arbitration in the air, reduction of armaments and the submission of points of difference to referees. Hence every action which tends towards amity between nations is to be commended, and what more natural than that the united enemies of a common foe, the allied forces of intelligent and scientific action against, not a national but an universal peril should meet from time to time to suggest and adopt the most comprehensive means of checking the same. And such united action of human fighters against an elemental antagonist will, it is devoutly to be hoped, tend towards the elimination of the national fighter; not because the latter is unnecessary to the life of a nation as at present constituted, but rather because peace is the greatest of blessings and will bring in its train scientific development along the most useful and beneficial of lines to the common weal.

To the youth of all nations, the writer would make the following appeal. The career of the fire-fighter is one of the most enthralling that the mind of man can conceive and in its present stage of development it promises a remarkable field for the enterprising and enthusiastic. It has been shown that the days when the fireman was merely an automaton using a pail of water and a hatchet, when discretion and intelligence were useless owing to the undeveloped state of the science and when any unskilled labourer could accomplish all that was required, are gone forever. In place is a calling which is emerging from its chrysalis of obscurity to take its proper position amongst the recognized and esteemed professions of the world.

While, as has been pointed out in the preceding pages, it does not offer the financial returns or popular recognition attached to other occupations, it does promise a sufficiency. If followed with determination, positions of responsibility are within reach, and what the future holds for the "fireman" no one can foresee. More unlikely things have occurred in the evolution of society than the formation of a national force of firemen, paid and recruited by the government and having in their charge the fire risks of the community as a whole, worthy successors of the soldier and the sailor, if that time ever arrives when the sword shall be laid aside for the ploughshare. And who shall say that it will not! Which brings the writer to the **end** of his labours.

The day is closing in and the lights are beginning to twinkle across the harbour of New York, as though beckoning to the wanderer to lay aside the cares of the moment and find rest and safety within their embrace. And somehow there steals across the evening air a vague feeling of sadness, the consciousness that many hands generously extended in the past as tokens of friendship and guerdons for courage will never again be clasped; that the enemy, "fire," has reaped a rich harvest, and that in dealing in pen and ink with the antagonist of a lifetime, in the words of that great empire builder, Cecil Rhodes, there has been "so much to do, so little done." But none the less, from the writer's eyrie overlooking the restless Hudson, bearing upon its broad bosom the commerce of the seven seas, he holds out his hand in greeting to the new recruit, in friendship to the active rank and file and in congratulation to the veteran fire-fighters the world over; in all climes, in all cities, in all countries, the greatest brotherhood of the world for the common weal.

PRACTICAL TESTS FOR FIRE ENGINES

I⊤ is the purpose of this appendix to set forth convenient and practical methods of making fire engine tests which will show the physical condition of engines, their capacity for delivering water at a reasonable pressure and the ability of the operating crews. The method described has been in use for a number of years and has been found practical, exact and of great value. Although methods similar to that described below are in use in some departments, the character of tests made in many cities, and especially those for acceptance, are usually more spectacular than exact. The throwing of a stream over a church spire, city hall or court house does not necessarily show that the engine is capable of delivering its full rated capacity at a proper working pressure.

Investigation has shown that where regular and systematic tests of engines are not made, even in well managed fire departments, defects often exist which may continue unsuspected for considerable periods and become manifest under the stress of a large fire, where the engine is called upon to deliver its full capacity under suitable working pressures. Furthermore, regular tests are a most valuable drill for engine crews, for in only a few departments do they receive sufficient training in operating engines to capacity. The breakdown of an engine at a fire or the inability of the crew to operate it to capacity may be the direct cause of confusion and needless loss of property and perhaps of life, to the discredit of the department.

Contracts for new fire engines usually contain guarantees that the engine will deliver a certain quantity of water, but often do not specify the pressure at which it is to be delivered, nor provide for any definite tests which will accurately determine whether the engine has fulfilled the guarantee; or, in other words, if the department is getting what it is paying for. In several cities, engines are required to fill large measured tanks in a specified time, but this is a cumbersome method at best, and such tanks are frequently unavailable; this usually gives no definite results as to pressure obtained and power developed.

STEAM FIRE ENGINE

A PRACTICAL test should show, with fair accuracy, the condition of both water and steam ends of pumps and the condition of the boiler; determine the amount of water which the engine will pump at a reasonable working pressure, such as would be required when operating at a large fire; demonstrate the ability of the engine to draft water, whether the pumps and waterways are tight under high pressures and steam valves are properly set, and whether the coal used is quick steaming and free from objectionable impurities. In addition, the test should be of such a character as to approach the working condition at a serious fire where the full capacity of the engine would be required, and at the same time be easily understood. The following tests are intended to bring out all of these points.

The displacement test indicates very closely the actual condition of the pumps as a whole and, in conjunction with the high pressure and valve tests, the condition of the plungers, pump valves, packing, etc. The high pressure test, in connection with the results obtained from the capacity test, indicates the setting of steam valves and condition of steam cylinders and packed joints. The capacity test shows the steaming quality of the boiler under heavy draft and the ability of the engine to make sufficient speed to develop its capacity when working against a reasonable water pressure. If the test is made from a cistern or reservoir, it will show the ability of the engine to draft; if made from a hydrant, the percentage of slip obtained will indicate this feature, as an engine showing less than 5 per cent. slip may be depended upon to take suction satisfactorily. Incidentally, the test also shows the ability of the engine crew in operating and stoking the engine.

Any machine, when new, should be capable of greater work than after several years of service; for this reason, a new engine should be given an acceptance test at least as severe as any work it may have to perform in actual service. This test should bring out not only the capacity to pump the actual volume of water specified by the maker as the rated capacity, but also to do this at a good working pressure.

A good specification, applying equally well to steam fire engines and automobile fire engines, is that the engine should deliver its full rated capacity at 120 pounds net pressure and 50 per cent. of its rated capacity at 200 pounds net pressure. This will assure sufficient boiler capacity in steam fire engines and gasoline engines of high enough power in the automobile fire engine.

Engines in service need not be given as severe a test as those being accepted, as it is mainly their general condition that is to be ascertained; for this reason, 100 pounds net water pressure would seem a sufficiently high requirement for the ordinary capacity test, which should be made at least yearly.

Apparatus Necessary for Testing.—For the tests outlined below, no elaborate or costly outfit is needed, the only special appliances absolutely required being as shown on Plate I and listed below:

A revolution counter. (Figure 3.)
A stop-watch and wrist strap. (Figure 5.)
A small Pitot tube. (Figure 9.)
An air chamber on Pitot. (Figure 11.)
Two or more pressure gages. (Figures 1 and 10.)
A set of smooth bore nozzles. (Figure 4.)
A hydrant or engine-discharge cap. (Figure 2.)
Appliance for attaching counter. (Figures 6 and 7 or
    Figure 8.)

The revolution counter should be of a type easily attached to the engine frame, or any convenient part, and so made as to register accurately at any speed likely to be reached by a reciprocating engine and be easily read.

The counter may be provided with straps for attaching to engine, or with the clamp and angle iron shown on Plate I, Figures 6 and 7, or with bolts and slotted lugs as shown in Plate I, Figure 8.

Tachometers and speed indicators are unsuitable for fire engine work, as the vibration is apt to render their readings unreliable.

A stop-watch can be purchased for less than $10, although an ordinary watch can be used.

The Pitot tube may be any of several suitable types now on the market, or the type shown on Plate I may be readily constructed. Dimensions are given below. It should be connected by ¼-inch brass pipe fittings to a pressure gage; to prevent vibration of the needle, an air chamber should be provided as shown on Plate I.

The pressure gages should be preferably not more than 3½ inches in diameter, in order that they may be conveniently handled. They should be of the compound type, in order that any disarrangement of the needle may be readily observed, one capable of indicating from a vacuum up to 150 pounds and one up to 200 pounds, and preferably divided for every pound and marked every 5 or 10 pounds, as shown in Figures 1 and 10, Plate I. Gages, especially those used with the Pitot, should be of good quality and accurate. They should be carefully calibrated (tested) with a weight tester or a standard gage before each day's work.

Nozzles suitable for testing are usually found in the regular equipment of every fire department. Only smooth bore tapered nozzles should be used, as discharges from ring nozzles are uncertain. Care should be taken that the tips are not nicked or otherwise injured, and that washers do not project into the pipe, as a perfectly smooth waterway is essential. The ring nozzles on many engines have

TESTING KIT.

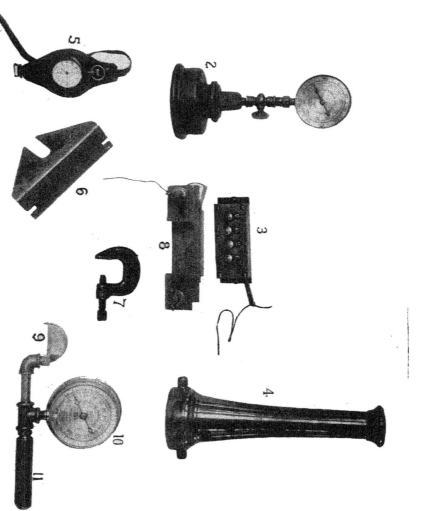

loose rings, which may be slipped out by unscrewing the end cap, leaving a suitable smooth-bore tip. Shut-off nozzles should not be used, as these generally have interior projections or breaks in the waterway, likely to cause eddies in the stream. Where much testing is to be done, it is better to set aside nozzles, keeping them solely for that purpose. The bore of nozzles should be accurate to size within 1/1,000 of an inch and carefully measured.

The engine-discharge cap, or hydrant cap (in most cities these have the same thread) is tapped for ¼-inch pipe thread and fitted with a nipple and stop-cock for attaching the test gage. By attaching to the discharge outlet of the engine, the engine water gage and the test gage may be compared to determine if the engine gage is correct. Where there is time to detach the water gage and a testing set is available, the gage can be more accurately checked. The steam gages are less likely to get out of order, being less subject to sudden fluctuations, and a comparison of readings of side and rear steam gages will usually be sufficient. If the engine has no suction gage or tapped suction cap, the engine or hydrant cap should be used on the second outlet of the hydrant when testing an engine at a double outlet hydrant.

Tests are best made by a supervisor (as the master mechanic or other officer conducting the test will hereafter be called), with an assistant accustomed to reading gages. Tables showing the discharge at various pressures through different nozzles, for use with Pitot tube readings, are to be found on pages 382 and 383. The suitable form for recording data of tests is shown on page 372, and until the supervisor becomes familiar with tests, it is advisable to use a similar form at the tests in order not to overlook any necessary data. Later, a pocket note-book will doubtless be found more convenient, care being taken to record all the necessary data.

Preliminary to Test.—If possible, calibrate gages of engine before the test, by detaching and comparing on a portable gage-testing set. They should be calibrated in the position in which they are to be used, either horizontally or vertically. If this is not done, check water and suction gages at test, as explained below.

If it is desired to determine the ability of the regular engine crew, the engine should, of course, be operated by them; if the condition and capacity of the engine are the unknown factors, a crew known to be efficient should be selected.

If there is any convenient body of water, or cistern, where water may be drafted with not over 10 feet of lift, then test should be made at draft; otherwise, attach engine to hydrant, care being taken to get a hydrant attached to a large main (8-inch or larger), and that the hydrant pressure is not excessive, preferably below 40 pounds. Four-inch or larger suction should be used. After suitably stationing engine, light the fire; note the time when smoke comes from stack, when the steam gage needle moves, at 50 pounds of steam, at 100 pounds, and pressure and time of blowing off. If engine has hot water in boiler, this may be omitted, noting only the pressure at which safety valve blows off. Then, if water gage on engine has not been calibrated (checked), attach hydrant cap and 200-pound test gage to engine discharge outlet. Record zero of all three gages—water, suction and test gages; open hydrant and record static pressure on all three gages; then with churn (hand relief) valve partly open and discharge gates shut, pump up pressure and compare test and water gages at 80 pounds, 100, 110, 120, etc., up to 120 pounds over the static or hydrant pressure. If engine has no suction gage, one of the suction caps on the engine can be tapped to connect the gage, or the engine or hydrant cap provided with the second gage should be attached to one hydrant outlet.

Let supervisor and assistant compare watches and set second hands together, or nearly so; this is more quickly accomplished if one watch has a stop-hand. The supervisor will find it convenient to tie his watch to coat or wrist in order to leave his hands free to hold note-book or Pitot. A leather watch holder and wrist strap, as shown on Plate

I, such as any harness maker can make, is a convenient appliance for this purpose. Attach the revolution counter and connect with one of the eccentric strap oil cups or studs by a short length of cord; have engine started slowly and adjust counter cord so that each revolution registers.

Displacement and Capacity Test.—While the engine is getting up steam, have firemen lay hose and connect nozzle. If testing on a paved street, it is best to lay nozzle down in gutter. Use a play-pipe holder or tie nozzle to any convenient post, in order to prevent pipe getting away from pipeman and doing damage.

For the larger engines, attach a line of hose on each side of the engine and connect into the siamese of a deluge set.

With the smaller size engines, it is usually more convenient to use a single line from one side of the engine; when deluge sets are not available, single lines may be used on the larger engines. In the table on page 379, the length of hose and size of nozzle best adapted for testing engines of various sizes are given. In testing with the siamesed lines, start the engines with both lines open and bring it up to speed; if the desired water pressure is not obtained, close the discharge gate on one line slowly until the gate indicates the proper pressure. Similarly, with a single line attached, the gate is closed slowly after engine has obtained its full speed until the desired pressure is obtained.

The supervisor can, from time to time, regulate this discharge gate to keep the desired water pressure, although if the crew operates the engine properly but little change will have to be made throughout the test. The engineer can be instructed to direct all his attention to operating his engine to full capacity, and the supervisor or testing engineer can regulate the water pressure, take the readings of the revolution counter, steam, water and suction gages, while his assistant takes readings of the nozzle pressure throughout the test.

When siamesed lines are used, should the engine not be able to maintain the desired water pressure with one line shut off entirely, add another length of hose to each side, or use a nozzle ⅛-inch smaller. With single lines, when the engine cannot maintain the desired pressure without undue throttling of the discharge valve, use a smaller nozzle or add another length of hose. The nozzle readings should, if possible, be over 40 pounds, as below this point readings must be very nearly constant to give accurate results.

Should water pressure at the engine be too high with both lines wide open, use a larger nozzle or cut out a length of hose from each side.

Relief valves should be closed, sprinkler used only as needed, and feed pumps operated regularly. The capacity test should last at least 20 minutes from the time the engine reaches full speed. During this time the water pressure at the engine should be constant and such as to give a net water pressure over the suction pressure of 100 to 120 pounds. Unless the rubber tires cause undue vibration, a modern engine, if in good condition, can safely run for an indefinite period at 400 to 425 feet of piston travel per minute, that is, 300 to 320 revolutions for an 8-inch stroke.

It is usually better to hold about 10 pounds over the pressure actually required, when the water pressure fluctuates much, as most engineers read the top of swing of a gage needle, while the supervisor, of course, should read the middle of the vibration. Gages may be throttled to prevent excessive vibration, but should always show some vibration to get true readings. A better method of preventing excessive vibration of needle on gage is to attach a small air chamber to the connection near the gage. During the capacity test, the supervisor should read counter (exactly at minute) and steam, water and suction gages each minute in regular order, and note the handling and stoking, feed water, leaks, uneven steam pressure, blowing off, foaming of boiler, accidents, and the other little details which his experience teaches him to observe. Meanwhile the supervisor's assistant should read the nozzle pressure every ¼ minute. Special care should be taken in reading the nozzle pressure. The Pitot should be held in the middle of the stream, with the tip about one-half the diameter of the bore from the end of the nozzle. Gage should be horizontal or vertical, according to the position in which it was calibrated, and at the same level as the end of the nozzle.

High Pressure Test.—After a run of 20 minutes in which there were no serious interruptions to readings, and pressure was maintained at an average of at least 100 pounds net, stop stoking; shut down, close discharge gates, partly open churn valve and get steam down to between 70 and 80 pounds, drawing fire if necessary. Then start engine slowly, and gradually close churn valve tight. See that all other openings, feed pumps, sprinklers, relief cocks, etc., are shut. Let engine turn in this condition for one or two minutes; observe the number of revolutions, and the water, steam and suction (now static) pressures; note any uneven motion of engine, blowing through of steam or imperfect valve setting, leaks in steam or water ends, or fittings, etc. If pumps are in good condition and valves set correctly, speed should not be over one revolution in 10 seconds in any modern type engine. With 70 pounds steam and 50 pounds suction, water pressure will reach about 250 pounds; this is perfectly safe and not a severe test, as such pressures are frequently met in operation when long lines are used.

Valve Tests.—After taking the observations for the high pressure test, shut off throttle of engine and open cylinder drips. Note the drop in water pressure for say one-half minute. The manner in which this pressure holds up is an indication of the condition of the discharge valves. A drop of not over 15 pounds in one-half minute, provided there are no external leaks visible around the pump, indicates a fairly good condition of the valves.

Suction Test.—If the engine has been tested at a hydrant, its ability to draft may be determined as follows, provided it is equipped with a compound suction gage or one of the suction caps is tapped to receive a compound gage: Disconnect engine from hydrant while there is still some steam pressure on boiler, put both suction caps on tight, open one of the discharge gates and then open throttle, allowing en-gine to run at a moderate speed, observe the reading of the compound gage while running, and also after shutting down. The drop of the vacuum after shutting down is an indication of the condition of the suction valves, provided all joints are good.

To Figure Displacement.—In averaging the nozzle, steam, water and suction pressures, subtract ½ of first and last readings from sum of readings used (see sample test sheet). Average the nozzle pressure during a period in which the engine ran steadily, water pressure was well maintained and the nozzle pressure varied the least. When possible, use a 20-minute period in figuring the dis-placement; if for any reason there is much variation in the nozzle pressure, say over 10 per cent. during any one min-ute, select as long a period as possible, but at least 10 minutes, during which the pressure has been well main-tained. Correct for gage error. Take out corresponding gallons from table, interpolating for odd pressures or for odd-size nozzles.

Example: 1½″ nozzle, 61 pounds nozzle pressure.

62 pounds nozzle pressure gives........ 525 gallons
60 pounds nozzle pressure gives........ 517 gallons
_____

or 2 pounds give a difference of......... 8 gallons
and 1 pound gives ½ of this, or........ 4 gallons
Therefore, 61 pounds nozzle pressure....—517+4
—521 gallons

Example: 1 9/16″ nozzle, 60 pounds nozzle pressure.

60 pounds through 1⅝″ nozzle gives..... 607 gallons
60 pounds through 1½″ nozzle gives..... 517 gallons
_____

or ⅛″ difference in nozzle diameter gives.. 90 gallons
and 1/16″ difference in nozzle dia'r gives 45 gallons
Therefore, 1 9/16″ nozzle at 60 pounds gives 517+45
—562 gallons

For odd-size nozzles, the discharge can be accurately ob-tained by using the formula under table of nozzle factors.

Divide the average gallons discharged by the average revolutions per minute to obtain the actual net displacement of the pumps. The nominal displacement will be found from the table, page 366, allowing for the pump rods. The dimension of the pumps, such as stroke, diameter of pump barrel and pump rods, should be accurately measured, if in question. The difference between actual and nominal displacements is the slip, which should be from 3 to 5 per cent. of the nominal displacement in a new engine (6 per cent. in a rotary); of this, about ½ per cent. is due to the feed water. After engine has been in use a few months, slip will generally increase about 1 per cent.; thereafter, if valves and packings are given proper attention, there should be only a slight increase. A slip of 10 per cent. or over indicates broken or displaced valve springs, and more than this, a badly worn plunger or pump barrel, or possibly a leaky suction. In a rotary, the wear is principally in the pump cam slides, which will also stick at times, causing increased slip even if not worn.

To Figure Capacity.—When the engine is run for 20 minutes at a uniform speed during the displacement test, the average discharge measured at the nozzle by the Pitot is the capacity of the engine. If only a 10-minute period of the run is used for figuring the displacement, the capacity of the engine is determined by multiplying the actual displacement (found in the displacement test) by the average revolutions per minute during a 20-minute period in which the engine worked at its full capacity. Steam, water and suction pressures during the capacity run should be averaged and corrected for gage error. In figuring percentage of capacity delivered, for a new fire engine, it is well to use contract figures for the rated capacity which the engine is guaranteed to deliver. A capacity due to a piston travel of about 420 feet per minute (315 revolutions for 8-inch stroke) less a 3 per cent. allowance for slip, is reasonable for a modern engine; older types vary considerably.

AUTOMOBILE AND GASOLINE DRIVEN FIRE ENGINES

IN so far as they apply, the same tests are desirable for automobile and other gasoline motor pumping engines as for steam fire engines, so that the same methods for measuring the water discharged, calibrating (testing) water gages, calculating the actual and normal displacement and slip of the pumps and averaging the net water pressure may be used. High pressure, valve and suction tests may also be made in much the same way as on steam fire engines.

Owing to the characteristics of the internal combustion engine certain additional tests and modifications will be found advantageous. A capacity test should be run longer than is usually necessary with a steam engine. It is suggested that for acceptance, engines of this type be required to deliver their full rated capacity at 120 pounds average net pressure for 2 hours, and 50 per cent. of their rated capacity at 200 pounds net pressure for 1 hour.

Additional tests with a line or lines at least 300 feet in length with shut-off nozzles are desirable and should preferably be made with engine drafting. While the streams are playing with 120 pound pressure at the engine, first one nozzle and then both should be shut off and the same tried with other pressures up to 200 pounds. In these tests the relief valve should be set at about 10 pounds higher than the pressure to be carried.

The pumps should be stopped and started with lines open, with one open and one closed, and with both closed; the motor should not stall during such tests. Tests should also be made to show that the engines can pump their full guaranteed capacity at 120 pounds net pressure when drafting with up to 10 feet of lift, if there is a possibility that the engine may be required to take suction from a river, canal or cistern when in service.

LOG OF FIRE ENGINE TEST

City._____ Engine No. 5 Make._____ Builder's No._____
Tested by A.C.B & K.H.C _____ Location Rear of quarters Date. 11-9-08
Run by Regular engineer Fired by Regular asst engineer Coal Jackson lump

GAGE COMPARISON TEST	WATER GAGE	NOZZLE PRESSURE				TIME	COUNTER	R.P.M.	STEAM	WATER	SUCTION	NOTES
GAGE NAME		Min.	⅛	½	¾							
0	1	80	81	80	81	339	7870		132	140	35	Engine at hydrant
90	91	80	82	80	81	40	8178	308	125	140	37	1 length 9' of 4" soft suction
100	101	82	84	87	86	41	8486	308	135	140	40	Boiler hot; fire lighted
120	122	88	88	90	92	42	8797	311	135	145	35	at 3.15, steam 50 at 328
140	144	90	93	93	92	43	9119	322	135	145	34	Small safety set at 130°
150	151	93	92	92	92	44	9442	333	132	145	35	Large safety set at 135°
160	161	91	91	91	90	45	9774	332	134	148	32	Hose ½in 2
		90	89	90	92	46	103	329	130	145	33	1 line 150'; 1 line 100'
60tr.		94	95	96	98	47	430	328	135	145	32	2½ C.R.L. Siamesed
60	62	97	96	93	92	48	759	328	135	145	32	to deluge set Nozzle 1¾
suction		90	90	87	84	49	1093	334	129	155	38	(1.6259) smooth bore
gage 67		82	86	83	87	50	1422	329	120	155	36	Check marks indicate
		86	87	86	87	50	1739	317	132	142	35	large safety valve opened
		92	93	96	97	52	2063	324	140	150	35	Crew O.K.
		85	95	94	95	53	2395	332	140	145	35	
		91	96	92	90	54	2728	333	132	155	35	
		87	86	90	90	55	3048	320	132	140	38	Engineers regular with feed
		92	87	91	95	56	3378	330	135	150	40	water
		94	92	92	88	57	3708	330	140	150	35	Stoker carries even fire
		86	88	92	93	58	4032	324	124	140	40	Luck satisfactory.
		90	1791	1785	1782	59	4351	326	135	145	35	Engine runs smoothly
		1870										
		85										
		1785										

DISPLACEMENT TEST		CAPACITY TEST		HIGH PRESSURE TEST					VALVE TEST	
				TIME	COUNTER	R.P.M.	STEAM	WATER	TIME	WATER
Time 339-59	+ 20 min	Duration mins	20	4.05	5228		72	220	4.07	130
Ave noz pres	89.7	Ave r pm	324	4.06	5230	2	72	200	4.08½	125
Corrections	+2	Gallons per min	750							
Corrected pres	91.7		DISPLACEMENT						SUCTION TEST	
Gallons per min	7498	Ave steam pres	133						TIME	SUCTION GAGE
Revs per min	324.4	Ave water pres	146/145½	REMARKS: No Scale					4.10	20"
Displacement	2311	Ave suct pres	356/36½						4.11	20"
" (normal)	2370	Net water pres	108½							
Slip per cent	3									

Reasonable capacity
of engine 70h gallons.
Obtained 750 gallons
or 107% of rating

Figured by A.C.B.
Checked by K.H.C.

A. L. A. M. FORMULA FOR HORSE-POWER OF GASOLINE MOTORS.

$$\text{Horse-Power} = \frac{\text{Bore} \times \text{Bore} \times \text{No. of Cylinders}}{2.5}$$

Example.—Six-cylinder motor 4½-inch bore.

$$\text{H.-P.} = \frac{4\frac{1}{2} \times 4\frac{1}{2} \times 6}{2.5} = 48.6$$

REASONABLE CAPACITIES OF MODERN FIRE ENGINES

Bore of Pumps, Inches	Stroke, Inches	Capacity, Gallons per Minute
6	9	1,100
5¾	8 or 9	1,000
5½	8	900
5¼	8 or 9	850
5	8	750
4¾	8	700
4½	7 or 8	600
4¼	7 or 8	550
4	7	500

RATED CAPACITY OF SILSBY ENGINES

Maker's Size	Nominal Displacement per Revolution, Gallons	Rated Capacity, Gallons per Min.
Extra First	1.261	1,000
First	1.141	900
Second	0.952	700
Third	0.804	600
Fourth	0.675	500
Fifth	0.513	400

CALCULATIONS FOR ENGINE TESTS.

DISPLACEMENT TEST.

AVERAGE DISCHARGE.

To obtain Average Nozzle Pressure:

Sum Column "Min."..........	1,870
Subtract ½ sum of first and last figures.....................	85
	1,785
Sum Column "¼".............	1,791
Sum Column "½".............	1,795
Sum Column "¾".............	1,802
Divide by 80...............)	7,173

Average Nozzle Reading.....	89.7
Correction from Gage Test Sheet.......................	+2.0
Average Nozzle Pressure...	91.7

From Discharge Tables for 1⅝" Nozzle:

92 lbs. gives...............751 gallons.
90 lbs. gives...............743 gallons.

2 lbs. gives...............8 gallons.
1.7 lbs. gives.............6.8 gallons.
Then 91.7 lbs. = 749.8 gallons.

AVERAGE R. P. M.

Counter at 3.59...............	4,358
Counter at 3.39...............	7,870
Divide by 20..............)	6,488
Average R. P. M. =	324.4

ACTUAL DISPLACEMENT.

$$\frac{\text{Average Discharge}}{\text{Average R. P. M.}} = \frac{749.8}{324.4} = 2.311$$

NOMINAL DISPLACEMENT.

From Engine Displacement Table:

4¾" Bore, 8" Stroke...........	2.455
1¼" Pump Rod...............	.085
Nominal Displacement =	2.370

SLIP, IN PER CENT.

$$\frac{\text{Nom. Displacement} - \text{Act. Displacement}}{\text{Nominal Displacement}}$$

$$\frac{2.370 - 2.311}{2.370} = 3\%.$$

CAPACITY TEST.

AVERAGE R. P. M.

Same as for Displacement Test in this case.

GALLONS PER MINUTE.

Same as for Displacement Test in this case.

AVERAGES OF PRESSURES.

Steam:

Sum of Column................	2,787
½ of first and last figures.......	133
Divide by 20.............)	2,654
Average Steam Reading....	132.7

Water:

Sum of Column..............	3,065
½ of first and last figures....	142.5
Divided by 20..........)	2,922.5
Average Reading.........	146.1
Correction from Test of Gage and Test Sheet, for Gage No. 119...................	—1.0
Average Water Pressure..	145.1

Suction:

Sum of Column................	746
½ of first and last figures.......	35
Divide by 20.............)	711
Average Reading...........	35.6
Correction from Test of Gage	+1.0
Average Suction Pressure...	36.6

Net Pressure:

Average water pressure........	145.1
Average suction pressure........	36.6
Average net pressure.......	108.5

PERCENTAGE OF CAPACITY OBTAINED.

Reasonable capacity of Pumps based on 400 ft. Piston Travel per Min...................	=700 gals.
Obtained at Test.............	750 gals.
or 107% of Rating.	

ENGINE DISPLACEMENT TABLE

DOUBLE PUMPS.

Bore of Pump Inches	PLUNGER DISPLACEMENT. GALLONS PER REVOLUTION.			Diameter of Pump Rods	PUMP ROD CORRECTION. GALLONS PER REVOLUTION.		
	Stroke in Inches				Stroke in Inches		
	7	8	9		7	8	9
3 1/2	1.166	1.333	1.500	1"	0.047	0.054	0.061
3 5/8	1.251	1.430	1.609	1 1/16	0.053	0.061	0.069
3 3/4	1.339	1.530	1.721	1 1/8	0.060	0.069	0.078
3 7/8	1.430	1.634	1.838	1 3/16	0.067	0.077	0.087
4	1.523	1.740	1.958	1 1/4	0.074	0.085	0.096
4 1/8	1.620	1.851	2.082	1 5/16	0.081	0.093	0.105
4 1/4	1.719	1.965	2.211	1 3/8	0.089	0.102	0.115
4 3/8	1.822	2.083	2.343	1 7/16	0.098	0.112	0.126
4 1/2	1.928	2.203	2.478	1 1/2	0.107	0.122	0.138
4 5/8	2.036	2.327	2.618	1 9/16	0.116	0.133	0.150
4 3/4	2.148	2.455	2.762	1 5/8	0.126	0.143	0.162
4 7/8	2.263	2.586	2.909	1 11/16	0.136	0.155	0.174
5	2.380	2.720	3.060	1 3/4	0.146	0.167	0.188
5 1/8	2.500	2.858	3.215				
5 1/4	2.624	2.999	3.374				
5 3/8	2.750	3.143	3.536				
5 1/2	2.880	3.291	3.702				
5 5/8	3.012	3.442	3.872				
5 3/4	3.147	3.597	4.047				
5 7/8	3.286	3.755	4.225				
6	3.427	3.917	4.407				

Subtract pump rod correction from plunger displacement to obtain correct displacement of engine.

For single-pump engines, use one-half of result obtained.

For single-acting pumps do not subtract pump rod connection.

Example: Engine with 5¼-inch pump, 9-inch stroke and 1½-inch pump rod. From Table above:

Displacement of Plunger = 3.374 gallons.
Correction for Rod = 0.138 gallons.

Nominal Displacement = 3.236 gallons.

Below is given a table for use when engines are worked at draft, either in actual service or in testing. A study of it will show that where a high lift is necessary, small suctions will restrict the capacity of an engine; the table indicates clearly what sizes are necessary under different conditions. The figures are based on the ability of the pumps to maintain a vacuum of 23 inches.

TABLE SHOWING MAXIMUM LIFT IN FEET WHEN DRAFTING VARIOUS QUANTITIES OF WATER WITH A FIRE ENGINE IN GOOD CONDITION.

Quantity of Water, Gallons per Minute	Maximum Lift in Feet, Engine Drafting				
	3″ Suction	3½″ Suction	4″ Suction	4½″ Suction	5″ Suction
300	16	20	22½	24	24½
400	8½	17	20	22½	24
500		12½	18½	20½	23
600		7	15	19½	21
700		4½	11	17	19½
800			6½	14½	19
900			6	11½	17
1,000				8	14½
1,100				7½	12
1,200				4	9½
1,300					6½
1,300	1	length of	suction.		9½

(right margin notes: 3 lengths of suction. / 2 lengths of suction.)

TABLE OF HOSE AND NOZZLES FOR TESTING ENGINE, USING SIAMESED LINES.

NOTE.—Connect Lines to a Deluge Set Provided with a Short Lead of 3½- or 4-inch Hose. Use Only Smth-Bore Nozzle and of the Diameter Given. By Regulating One of the Discharge Gates, Pressure can be Kept Nearly Constant and from Three-quarters to Full Capacity Obtained.

Size	Bore of Pump	Reasonable Capacity, Gallons per Minute*	Number and Length of Lines and Size of Nozzle Needed to Deliver the Reasonable Capacity at the Desired Pressure at the Engine			
			100 Pounds	120 Pounds	140 Pounds	160 Pounds
Double Extra First	6"	1,100	2-50' lines of 3" or 3-50' lines of 2½" 2" Nozzle	- 00' lines of 3" or 3-10' lines of 2½" 2" Nozzle	2-150' lines of 3" or 3-150' lines of 2½" 2" Nozzle	2-200' lines of 3" or 3-200' lines of 2½" 2" Nozzle
Extra First	5¾"	1,000	2-50' lines of 2½" 2" Nozzle	1-100' line of 2½" and 1-50' line of 2½" 2" Nozzle	2-100' lines of 3" 2" Nozzle	2-150' lines of 2½" 2" Nozzle
First	5½"	900	2-50' lines of 2½" 1⅞" or 2" Nozzle	1-100' line of 2½" and 1-50' line of 1⅞" Nozzle	2-100' lines of 2½" 1⅞" Nozzle	2-150' lines of 2½" 1⅞" Nozzle
Second	5"	750	2-50' lines of 2½" 1¾" Nozzle	2-100' lines of 2½" 1¾" Nozzle	2-150' lines of 2½" 1¾" Nozzle	2-250' lines of 2½" 1¾" Nozzle
	4¾"	700	2-50' lines of 2½" 1⅝" Nozzle	2-100' lines of 2½" 1⅝" Nozzle	2-150' lines of 2½" 1⅝" Nozzle	2-250' lines of 2½" 1⅝" Nozzle
Third	4½" or 4¼"	600 or 550	1 -0' line of 2½" and 1-150' line of 2½" 1⅝" Nozzle	1-100' line of 2½" and 1-150' line of 2½" 1½" Nozzle	-0' line of 2½" and 1-200' line of 2½" 1½" Nozzle	2-250' lines of 2½" 1½" Nozzle
Fourth	4"	500	2-100' lines of 2½" 1⅜" Nozzle	2-200' lines of 2½" 1⅜" Nozzle	2-300' lines of 2½" 1⅜" Nozzle	1-100' line of 2½" 1⅜" Nozzle †
Fifth	3⅝"	400	2-100' lines of 2½" 1¼" Nozzle	1-100' line of 2½" 1¼" Nozzle †	1-150' line of 2½" 1¼" Nozzle †	1-200' line of 2½" 1¼" Nozzle †

* Based on about 400' piston travel per minute. † Single lines; deluge set omitted.

NOTE.—If hose has not smoothest lining, shorter lines or a larger nozzle may be required; if hose is slightly larger than given on page 375, it may be necessary to use longer lines or a smaller nozzle.

TABLE OF ... ZES FOR ... ING ENGINES, ... ING SINGLE 50-FOOT LINES OF HOSE.

NOTE.—Connect ... ine to Nozzle; Bring Engine to Speed and Regulate Discharge ... as; if Desired Pressure Cannot be ... ed, Use ... the ⅛" Smaller or Add Another Length of ... e.

Size	Bore of Pump	Reasonable ..., Gallons per Minute*	Size of Nozzle Needed to Deliver the Reasonable Capacity at the Engine Desired Pressure at the Engine			
			100 Pounds	120 Pounds	140 Pounds	160 Pounds
First	5½"	900	2¼"	Single 50-foot Line of 3" hose 2¼" or 2"	2"	1⅞" or 1¾"
Second	5", 4¾"	750, 90	2", 1⅞"	Single 50-foot Line of 2½" hose 1⅞", 1¾"	1⅞", 1¾"	1⅝", 1⅝"
Third	4½", 4¼"	600, 550	1¾" or 1⅝", 1⅝"	1⅝" or 1½", 1½"	1½", 1⅜"	1⅜", 1⅜"
Fourth	4"	500	1½"	1⅜"	1⅜"	1¼"
Fifth	3⅝"	40	1¼"	1¼"	⅞"	1⅛"

* Based on a ... but 400' ... tion travel per mi ... ts.

FIRE STREAM TABLES

THESE tables are arranged to show the pressures required at the hydrant or fire engine, while stream is flowing, to maintain nozzle pressures given in the first columns, through various lengths of 2½, 3 and 3½-inch rubber lined hose in single lines and two lines of 2½-inch hose siamesed.

Nozzle pressures of 40 to 60 pounds from 1⅛ and 1¼-inch nozzles will give streams which may be classed as good and which can be handled without special appliances; for deluge sets, turret pipes, etc., with 1½-inch and larger nozzles, 80 to 120 pounds nozzle pressure is desirable for effective fire-fighting; the height, area and general character of the building are factors in determining at what pressure a stream may be considered good, as well as in determining whether a nozzle is of sufficient size to furnish an effective stream, nothing less than 1⅛-inch being considered as effective for outside work, except for fires in small buildings. In this connection it should be noted that a 1 or 1⅛-inch ring tip delivers a stream about ⅛ inch smaller than the diameter of the tip.

The pressure at the hydrant or fire engine is that indicated by a gage attached to the hydrant or fire engine while the stream is flowing. The pressure at the nozzle is that indicated by a Pitot gage held in the stream.

The hydrant (or engine) pressures are obtained by adding to the nozzle pressure the friction loss in the hose, and also the small additional loss in the hydrant outlet or engine discharge.

Friction losses in hose are based on tests of best quality rubber-lined fire hose and are for 100-foot lengths measured without pressure applied. Diameters of hose, as measured under 75 pounds pressure, assumed as the average working condition, were as follows: For nominal 2½-inch, 2.575 or about 2 9/16 inches; for nominal 3-inch, 3.125 or 3⅛ inches; for nominal 3½-inch, 3.685 or about 3 11/16 inches.

The smoothness of the lining has a very considerable effect on the friction loss, some samples tested showing losses 50 per cent. in excess of those given. A slight variation in diameter also produces a marked difference in friction loss; in the case of 2½-inch hose, a variation of 1/16 inch in diameter will result in 10 per cent. difference in loss. If properly beveled 2½-inch couplings are used on 3-inch hose, the loss of pressure due to them will be less than 5 per cent. of that gained by the use of the larger hose. For instance, for a flow of 300 gallons per minute, the loss in 2½-inch hose will be about 21 pounds, in 3-inch hose with 3-inch couplings about 8 pounds, and in 3-inch with 2½ inch couplings about 8½ pounds.

For siamesed lines, an allowance was made for the loss in the siamese connection and for 20 feet of 3½-inch lead hose.

The pressures given are for the nozzle at the same elevation as the hydrant or engine discharge outlet. Add or subtract 1 pound to the pressure given for each 2 1/3 feet difference in elevation. The arrangement of the table al lows a comparison to be readily made of the results obtain able with 3-inch hose and siamesed lines against single lines of 2½-inch hose.

TABLE OF NOZZLE FACTORS

THE discharge in gallons per minute is equal to the square root of the pressure multiplied by the factor.

Diameter of the nozzle in inches	FACTORS	
	For Fresh Water	For Salt (sea) Water
2	118.96	117.45
2¼	150.56	148.64
2½	185.88	183.50
2¾	224.91	222.05
3	267.66	264.25
3¼	314.13	310.13
3½	364.32	359.68
3¾	418.23	412.90
4	475.85	469.79
4¼	537.19	530.35
4½	602.25	594.58
4¾	671.02	662.48
5	743.51	734.03
6	1,070.64	1,057.00

For any size nozzles, the discharge, for fresh water, can be determined by the following formula:

Gallons per minute $= 29.83 \; c \; d^2 \; \sqrt{p}$.

Where d=diameter of nozzle in inches, measured to 1/1000 of an inch.

p=pressure recorded on Pitot gage in pounds.

c=a constant, varying from 0.990 for 1-inch nozzle to 0.997 for 6-inch nozzle.

For ordinary use, the formula can be reduced to:

Gallons per minute $= 29.7 \; d^2 \; \sqrt{p}$.

RMULA FOR OBTAINING APPROXIMATE NOZZLE OR ENGINE
PRESSURES, LENGTH OF LINE AND SIZE OF
NOZZLE BEING GIVEN.

$$\text{Nozzle Pressure in pounds} = \frac{\text{Engine Pressure}}{1.1 + K\,L}$$

Engine Pressure in pounds = Nozzle Pressure $(1.1 + L)$.

L = Number of 50-foot lengths of hose.

K = Constant, varying with size of nozzle and hose. See Table following.

Size Nozzle, Inches	K FOR					
	Single Line 2½" Hose	Single Line 3" Hose	Single Line 3½" Hose	Two 2½" Lines Siamesed *	Two 3" Lines Siamesed *	3 Lines 2½" Hose *
1	.105	.038	....	.025	....	....
1⅛	.167	.062	....	.043	....	....
1¼	.248	.092	.039	.066	.023	.028
1⅜	.341	.137	.059	.096	.034	.043
1½	.505	.192	.084	.135	.051	.061
1⅝	.680	.266	.113	.184	.068	.084
1¾	.907	.351	.152	.242	.093	.115
2	1.550	.605	.250	.418	.157	.190

* Allowance is made for loss in deluge set; these values will also give approximately correct figures for turret nozzles and water tower, except that in the latter, pressure equal to 0.434 times the height of tower must be subtracted from the engine pressure, before solving for nozzle pressure.

EFFECTIVE REACH OF FIRE STREAMS.

SHOWING THE DISTANCE IN FEET FROM THE NOZZLE AT WHICH STREAMS WILL DO EFFECTIVE WORK WITH A MODERATE WIND BLOWING. WITH A STRONG WIND THE REACH IS GREATLY REDUCED.

Pressure at Nozzle.	Size of Nozzle.									
	1-Inch		1⅛-Inch		1¼-Inch		1⅜-Inch		1½-Inch	
	Vertical Distance, Feet	Horizontal Distance, Feet	Vertical Distance, Feet	Horizontal Distance, Feet	Vertical Distance, Feet	Horizontal Distance, Feet	Vertical Distance, Feet	Horizontal Distance, Feet	Vertical Distance, Feet	Horizontal Distance, Feet
20	35	37	36	38	36	39	36	40	37	42
25	43	42	44	44	45	46	45	47	46	49
30	51	47	52	50	52	52	53	54	54	56
35	58	51	59	54	59	58	60	59	62	62
40	64	55	65	59	65	62	66	64	69	66
45	69	58	70	63	70	66	72	68	74	71
50	73	61	75	66	75	69	77	72	79	75
55	76	64	79	69	80	72	81	75	83	78
60	79	67	83	72	84	75	85	77	87	80
65	82	70	86	75	87	78	88	79	90	82
70	85	72	88	77	90	80	91	82	92	84
75	87	74	90	79	92	82	93	84	94	86
80	89	76	92	81	94	84	95	86	96	88
85	91	78	94	83	96	87	97	88	98	90
90	92	80	96	85	98	89	99	90	100	91

NOTE.—Nozzle pressures are as indicated by Pitot tube. The horizontal and vertical distances are based on experiments by Mr. John R. Freeman, *Transactions*, Am. Soc. C. E., Vol. XXI.

FRICTION LOSS IN FIRE HOSE.

BASED ON TESTS OF BEST QUALITY RUBBER LINED FIRE HOSE.*

Flow, Gallons per Minute	Pressure Loss in Each 100 Feet of Hose, Pounds per Sq. Inch				Flow, Gallons per Minute	Pressure Loss in Each 100 Feet of Hose, Pounds per Sq. Inch		
	2½" Hose	3" Hose	3½" Hose	2 Lines of 2½" Siamesed		3" Hose	3½" Hose	2 Lines of 2½" Siamesed
140	5.2	2.0	0.9	1.4	525	23.2	10.5	16.6
160	6.6	2.6	1.2	1.9	550	25.2	11.4	18.1
180	8.3	3.2	1.5	2.3	575	27.5	12.4	19.0
200	10.1	3.9	1.8	2.8	600	29.9	13.4	21.2
220	12.0	4.2	2.1	3.3	625	32.0	14.4	23.0
240	14.1	5.4	2.5	3.9	650	34.5	15.5	24.8
260	16.4	6.3	2.9	4.5	675	37.0	16.6	26.5
280	18.7	7.2	3.3	5.2	700	39.5	17.7	28.3
300	21.2	8.2	3.7	5.9	725	42.3	18.9	30.2
320	23.8	9.3	4.2	6.6	750	45.0	20.1	32.2
340	26.9	10.5	4.7	7.4	775	47.8	21.4	34.2
360	30.0	11.5	5.2	8.3	800	50.5	22.7	36.2
380	33.0	12.8	5.8	9.2	825	53.5	24.0	38.4
400	36.2	14.1	6.3	10.1	850	56.5	25.4	40.7
425	40.8	15.7	7.0	11.3	875	59.7	26.8	43.1
450	45.2	17.5	7.9	12.5	900	63.0	28.2	45.2
475	50.0	19.3	8.7	13.8	1,000	76.5	34.3	55.0
500	55.0	21.2	9.5	15.2	1,100	91.5	41.0	65.5

* Rough rubber lining is liable to increase the losses given in the table as much as 50 per cent.

DISCHARGE TABLE FOR SMOOTH NOZZLES.

NOZZLE PRESSURE MEASURED BY PITOT GAGE.

Nozzle Pressure in lbs. per sq. inch	NOZZLE DIAM. IN INCHES					Nozzle Pressure in lbs. per sq. inch	NOZZLE DIAM. IN INCHES				
	1	1⅛	1¼	1⅜	1½		1	1⅛	1¼	1⅜	1½
	Gallons per minute						Gallons per minute				
5	66	84	103	125	149	60	229	290	357	434	517
6	72	92	113	137	163	62	233	295	363	441	525
7	78	99	122	148	176	64	237	299	369	448	533
8	84	106	131	158	188	66	240	304	375	455	542
9	89	112	139	168	200	68	244	308	381	462	550
10	93	118	146	177	211	70	247	313	386	469	558
12	102	130	160	194	231	72	251	318	391	475	566
14	110	140	173	210	249	74	254	322	397	482	574
16	118	150	185	224	267	76	258	326	402	488	582
18	125	159	196	237	283	78	261	330	407	494	589
20	132	167	206	250	298	80	264	335	413	500	596
22	139	175	216	263	313	82	268	339	418	507	604
24	145	183	226	275	327	84	271	343	423	513	611
26	151	191	235	286	340	86	274	347	428	519	618
28	157	198	244	297	353	88	277	351	433	525	626
30	162	205	253	307	365	90	280	355	438	531	633
32	167	212	261	317	377	92	283	359	443	537	640
34	172	218	269	327	389	94	286	363	447	543	647
36	177	224	277	336	400	96	289	367	452	549	654
38	182	231	285	345	411	98	292	370	456	554	660
40	187	237	292	354	422	100	295	374	461	560	667
42	192	243	299	363	432	105	303	383	473	574	683
44	196	248	306	372	442	110	310	392	484	588	699
46	200	254	313	380	452	115	317	401	495	600	715
48	205	259	320	388	462	120	324	410	505	613	730
50	209	265	326	396	472	125	331	418	516	626	745
52	213	270	333	404	481	130	337	427	526	638	760
54	217	275	339	412	490	135	343	435	536	650	775
56	221	280	345	419	499	140	350	443	546	662	789
58	225	285	351	426	508	145	356	450	556	674	803
60	229	290	357	434	517	150	362	458	565	686	817

Assumed coefficient of discharge per cent. =.99   .99   .99   .99¼   .99½

NOTE.—Coefficients of discharge are based on experiments by Mr. John R. Freeman, *Transactions* Am. Soc. C. E., Vols. XXI and XXIV.

DISCHARGE TABLE FOR SMOOTH NOZZLES.
NOZZLE PRESSURE MEASURED BY PITOT GAGE.

Nozzle Pressure in lbs. per sq. inch	Nozzle Diam. in Inches					Nozzle Pressure in lbs. per sq. inch	Nozzle Diam. in Inches				
	1⅝	1¾	1⅞	2	2¼		1⅝	1¾	1⅞	2	2¼
	Gallons per minute						Gallons per minute				
5	175	203	234	266	337	60	607	704	810	920	1167
6	192	223	256	292	369	62	617	716	823	936	1187
7	207	241	277	315	399	64	627	727	836	951	1206
8	222	257	296	336	427	66	636	738	850	965	1224
9	235	273	314	357	452	68	646	750	862	980	1242
10	248	288	330	376	477	70	655	761	875	994	1260
12	271	315	362	412	522	72	665	771	887	1008	1278
14	293	340	391	445	564	74	674	782	900	1023	1296
16	313	364	418	475	603	76	683	792	911	1036	1313
18	332	386	444	504	640	78	692	803	924	1050	1330
20	350	407	468	532	674	80	700	813	935	1063	1347
22	367	427	490	557	707	82	709	823	946	1076	1364
24	384	446	512	582	739	84	718	833	959	1089	1380
26	400	464	533	606	769	86	726	843	970	1102	1396
28	415	481	554	629	799	88	735	853	981	1115	1412
30	429	498	572	651	826	90	743	862	992	1128	1429
32	443	514	591	673	854	92	751	872	1002	1140	1445
34	457	530	610	693	880	94	759	881	1012	1152	1460
36	470	546	627	713	905	96	767	890	1022	1164	1476
38	483	561	645	733	930	98	775	900	1032	1176	1491
40	496	575	661	752	954	100	783	909	1043	1189	1506
42	508	589	678	770	978	105	803	932	1070	1218	1542
44	520	603	694	788	1000	110	822	954	1095	1247	1579
46	531	617	710	806	1021	115	840	975	1120	1275	1615
48	543	630	725	824	1043	120	858	996	1144	1303	1649
50	554	643	740	841	1065	125	876	1016	1168	1329	1683
52	565	656	754	857	1087	130	893	1036	1191	1356	1717
54	576	668	769	873	1108	135	910	1056	1213	1382	1750
56	586	680	782	889	1129	140	927	1076	1235	1407	1780
58	596	692	796	905	1149	145	944	1095	1257	1432	1812
60	607	704	810	920	1168	150	960	1114	1279	1456	1843

Assumed coefficient of discharge per cent. = .995 .995 .996 .997 .997

1-INCH SMOOTH NOZZLE.—

Nozzle Pressure Indicated by Pitot Gage	Discharge, Gallons per Minute	Pressures Required at Hydrant or Maintain Nozzle Pressures given Lengths of Best Quality							
		Single 2½-inch Lines							
		100 Feet	200 Feet	300 Feet	400 Feet	500 Feet	600 Feet	700 Feet	800 Feet
20	132	25	30	35	39	44	49	53	58
25	148	31	37	43	49	55	60	66	72
30	162	38	44	51	58	65	72	78	85
35	175	44	52	59	67	75	83	91	98
40	187	50	59	68	77	86	94	103	112
45	198	56	66	76	86	96	106	115	125
50	209	62	73	84	95	106	117	128	139
55	219	68	80	92	104	116	128	140	152
60	229	75	88	101	114	127	140	153	166
65	238	81	95	109	123	137	151	165	179
70	247	87	102	117	132	147	162	177	192
75	256	93	109	125	141	157	173	189	205
80	264	99	116	133	150	167	183	200	217
85	272	105	123	141	159	177	195	212	230
90	280	111	130	149	167	186	205	224	243
95	287	117	137	157	177	196	216	236	256
100	295	123	144	165	185	206	227	247	268

2½- AND 3-INCH HOSE.

FIRE ENGINE, WHILE STREAM IS FLOWING, TO
IN FIRST COLUMN, THROUGH VARIOUS
2½- AND 3-INCH RUBBER LINED HOSE

		Single 3-inch Lines				Two 2½-inch Lines Siamesed			Nozzle Pressure Indicated by Pilot Gage
1,000 Feet	1,200 Feet	800 Feet	1,000 Feet	1,200 Feet	1,500 Feet	1,000 Feet	1,500 Feet	2,000 Feet	
68	77	35	39	42	48	33	40	46	20
84	95	43	48	52	59	41	49	57	25
99	112	52	57	62	70	49	59	68	30
114	130	60	66	72	81	57	68	79	35
130	148	68	75	82	92	65	78	90	40
145	165	77	84	92	103	72	86	99	45
160	182	85	93	102	114	80	95	110	50
175	199	93	102	112	125	88	105	121	55
192	218	102	112	122	137	96	114	132	60
207	235	110	121	131	148	103	122	141	65
222	252	118	130	141	159	111	132	152	70
237	269	127	139	151	170	120	142	164	75
251	285	135	148	161	181	128	151	175	80
266	302	143	156	170	191	135	159	184	85
280	.....	151	165	180	202	143	169	195	90
295	.....	158	173	189	211	150	177	204	95
310	.....	167	183	199	223	157	186	215	100

1⅛-INCH SMOOTH NOZZLE.—

Nozzle Pressure Indicated by Ptot Gage	Discharge, Gallons per Minute	PRESSURES REQUIRED AT HYDRANT OR FIRE NOZZLE PRESSURES GIVEN IN FIRST QUALITY 2½- AND									
		Single 2½-inch Lines									
		100 Feet	200 Feet	300 Feet	400 Feet	500 Feet	600 Feet	700 Feet	800 Feet	1,000 Feet	1,200 Feet
20	167	28	35	42	49	56	64	71	78	92	107
25	187	35	44	53	62	71	79	88	97	115	133
30	205	42	52	63	73	84	95	05	16	137	158
35	221	49	61	73	85	97	110	122	134	158	183
40	237	55	69	83	96	110	124	138	151	179	206
45	251	62	77	93	108	123	139	154	169	200	230
50	265	69	86	103	120	137	154	171	188	222	256
55	277	76	94	112	131	149	168	186	204	241	278
60	290	83	103	123	143	163	183	203	223	263	304
65	301	89	111	132	154	175	197	218	240	283	326
70	313	96	119	142	165	188	211	234	257	303	....
75	324	103	128	152	177	202	227	252	276	325	...
80	335	110	136	162	188	215	241	267	294	....	....
85	345	116	144	171	199	226	254	282	309	....	....
90	355	123	152	181	210	240	269	298	327	....	....
95	365	130	160	191	222	252	283	314	....	....	....
100	374	136	168	201	233	265	297	329	....	....	....

2¼-INCH AND 3-INCH HOSE.

Engine, while stream is flowing, to maintain Column, through various Lengths of Best 3-inch Rubber Lined Hose												Nozzle Pressure Indicated by Pitot Gage
Single 3-inch Lines							Two 2½-inch Lines Siamesed					
400 Feet	600 Feet	800 Feet	1,000 Feet	1,200 Feet	1,500 Feet	1,800 Feet	800 Feet	1,000 Feet	1,200 Feet	1,500 Feet	1,800 Feet	
32	37	43	48	54	62	71	38	42	46	53	60	20
40	46	53	60	67	77	87	45	50	55	63	70	25
47	55	63	71	79	91	103	53	59	65	74	82	30
55	65	74	83	93	07	121	62	69	76	86	96	35
63	73	84	95	05	21	137	70	78	86	97	08	40
70	82	94	106	18	35	153	79	87	95	08	21	45
78	91	104	117	30	50	169	88	98	107	21	35	50
86	100	114	128	142	164	185	96	107	117	132	147	55
93	109	124	139	155	178	201	105	116	127	143	160	60
101	117	134	151	167	192	217	114	126	138	156	174	65
108	126	144	162	180	206	233	122	135	148	167	186	70
116	135	154	173	192	221	249	130	144	157	178	198	75
124	144	165	185	206	236	267	138	153	167	189	210	80
131	153	174	195	217	249	281	147	163	178	201	224	85
139	161	184	207	229	263	297	156	172	188	212	237	90
146	170	194	218	242	277	313	164	181	198	224	249	95
154	178	203	228	253	290	....	172	190	208	235	261	100

1¼-INCH SMOOTH NOZZLE.—

Nozzle Pressure Indicated by Pilot Gage	Discharge, Gallons per Minute	Pressures Required at Hydrant or Fire Pressures given in First Column, 2½- and 3-inch									
		Single 2½-inch Lines									
		100 Feet	200 Feet	300 Feet	400 Feet	500 Feet	600 Feet	700 Feet	800 Feet	1,000 Feet	1,200 Feet
20	206	32	42	53	64	75	85	96	107	128	149
25	230	40	53	66	79	92	105	118	131	158	184
30	253	48	63	79	95	110	126	142	157	189	220
35	273	55	73	91	109	127	145	163	181	217	253
40	292	63	83	104	124	144	165	185	206	246	287
45	309	70	93	116	138	161	183	206	229	274	319
50	326	78	103	128	153	178	203	228	253	303	....
55	342	86	113	140	167	194	222	249	276	330	....
60	357	93	123	152	182	211	241	270	300	....	....
65	372	101	133	164	196	228	260	292	323	....	....
70	386	108	142	176	210	244	278	312	....	....	....
75	399	116	152	188	224	261	297	333	....	....	....
80	413	124	163	201	240	279	318	....	....	....	....
85	425	131	172	213	254	295	....	..»	....	....	....
90	438	139	182	225	269	312	....	....	....	....	....
95	449	46	191	236	282	327	....	....	....	.	....
100	461	53	201	248	295	....	....	....	....	.	....

2½- AND 3-INCH HOSE.

Engine, while stream is flowing, to maintain Nozzle through various Lengths of Best Quality Rubber Lined Hose

Single 3-inch Lines							Two 2½-inch Lines Siamesed						Nozzle Pressures Indicated by Pitot Gage
400 Ft.	600 Ft.	800 Ft.	1,000 Feet	1,200 Feet	1,500 Feet	1,800 Feet	600 Ft.	800 Ft.	1,000 Feet	1,200 Feet	1,500 Feet	1,800 Feet	
37	46	54	62	70	83	95	39	45	51	57	67	76	20
47	57	67	77	87	102	117	48	55	62	70	80	91	25
56	68	81	93	05	123	142	57	66	74	83	96	109	30
65	79	92	106	120	141	161	66	76	86	95	110	125	35
74	89	105	120	136	159	183	75	87	99	·110	127	144	40
83	100	117	135	152	178	204	84	96	109	121	140	158	45
91	111	130	149	168	197	226	93	107	121	135	155	176	50
100	121	142	163	184	216	247	102	117	132	147	169	192	55
109	132	155	178	201	235	270	111	128	144	160	185	210	60
118	143	167	192	217	254	291	120	137	155	173	199	225	65
127	154	180	206	233	272	....	129	147	166	185	213	241	70
136	164	192	220	248	290	....	137	157	177	197	227	257	75
145	175	205	235	265	....	....	147	169	190	212	244	276	80
153	184	216	247	279	....	....	156	179	201	224	258	292	85
162	195	228	261	295	....	....	165	189	213	237	273	309	90
170	205	240	275	....	....	....	173	198	223	248	286	323	95
179	215	252	288	....	....	....	182	208	235	261	300	....	100

1⅜-INCH SMOOTH NOZZLE.—

Nozzle Pressure Indicated by Ptot Gage	Discharge, Gallons per Minute	Pressures Required at Hydrant or Fire Nozzle Pressures given in First Quality 2½- and									
		Single 2½-inch Lines									
		100 Feet	200 Feet	300 Feet	400 Feet	500 Feet	600 Feet	700 Feet	800 Feet	200 Feet	400 Feet
20	250	37	52	68	83	98	113	128	144	34	45
25	280	46	64	83	102	121	139	158	177	41	56
30	307	55	77	99	121	144	166	188	210	50	67
35	331	64	89	15	140	166	191	217	242	58	78
40	354	73	02	31	160	189	218	247	276	67	89
45	376	81	14	46	178	211	243	275	307	74	99
50	396	90	25	61	196	222	257	293	328	82	109
55	415	99	37	76	215	254	292	331	....	90	121
60	434	107	149	191	233	276	318	....	....	98	131
65	451	116	161	206	251	297	....	....	....	06	141
70	469	125	173	222	270	319	....	....	....	14	152
75	485	134	185	237	289	....	....	....	....	22	162
80	500	142	196	251	305	....	....	....	....	30	172
85	516	151	209	267	325	....	....	....	....	38	183
90	531	159	220	281	....	....	....	....	....	46	194
95	546	168	232	297	....	....	....	....	....	53	203
100	560	177	244	312	....	....	....	....	....	62	215

2½- AND 3-INCH HOSE

ENGINE, WHILE STREAM IS FLOWING, TO MAINTAIN COLUMN, THROUGH VARIOUS LENGTHS OF BEST 3-INCH RUBBER LINED HOSE												Nozzle Pressure Ind- cated by Pitot Gage
Single 3-inch Lines					Two 2½-inch Lines Siamesed							
600 Ft.	800 Feet	1,000 Feet	1,200 Feet	1,500 Feet	400 Feet	600 Feet	800 Feet	1,000 Feet	1,200 Feet	1,500 Feet	1,800 Feet	
57	68	80	92	109	37	46	54	63	71	84	96	20
70	85	99	113	135	46	57	67	78	88	104	119	25
84	01	18	135	161	56	68	81	93	06	124	143	30
97	17	37	157	187	65	80	94	08	22	143	165	35
112	34	57	180	214	74	90	106	122	138	162	186	40
125	50	75	200	238	83	101	119	137	155	182	209	45
137	64	92	220	267	92	111	131	151	171	201	230	50
151	82	12	242	288	100	122	144	165	187	219	252	55
163	96	29	262	....	109	133	156	180	203	238	273	60
177	12	47	282	....	118	143	168	194	219	257	294	65
189	27	65	303	....	128	155	182	209	236	277	317	70
203	43	83	....	....	137	165	194	223	252	295	....	75
215	57	00	....	....	145	175	206	236	266	312	....	80
229	74	..	....	....	153	186	218	250	282	331	....	85
241	89	..	....	....	162	196	230	264	298	....	....	90
254	04	..	....	....	170	206	241	277	313	....	....	95
267	..	..	....	....	179	217	254	291	329	....	....	100

1½-INCH SMOOTH NOZZLE.—

Nozzle Pressure Indicated by Pitot Gage	Discharge, Gallons per Minute	PRESSURES REQUIRED AT HYDRANT OR FIRE NOZZLE PRESSURES GIVEN IN FIRST QUALITY 2½- AND										
		Single 2½-inch Lines								Single		
		100 Feet	200 Feet	300 Feet	400 Feet	500 Feet	600 Feet	700 Feet	800 Feet	200 Feet	400 Feet	600 Feet
20	298	44	65	86	107	128	149	170	191	39	55	71
25	333	54	80	106	132	158	184	210	236	48	68	88
30	365	65	95	126	157	188	219	250	280	58	81	105
35	394	75	10	145	181	216	251	287	322	67	94	122
40	422	85	26	166	206	246	286	327	....	76	07	139
45	447	96	41	185	230	275	320	....	....	85	20	155
50	472	06	55	205	254	304	....	....	....	95	33	171
55	494	16	70	224	278	332	....	....	....	04	45	187
60	517	26	84	242	301	....	....	....	....	13	58	203
65	537	36	98	261	324	....	....	....	....	22	70	218
70	558	46	13	281	....	....	....	....	....	31	83	235
75	578	56	28	299	....	....	....	....	....	40	96	251
80	596	66	42	318	....	....	....	....	....	49	08	267
85	614	76	57	337	....	....	....	....	....	58	20	282
90	633	87	72	....	....	....	....	....	....	67	33	298
95	650	97	86	....	....	....	....	....	....	76	45	314
100	667	07	00	....	....	....	....	....	....	85	57	....

2½- AND 3-INCH HOSE.

ENGINE, WHILE STREAM IS FLOWING, TO MAINTAIN COLUMN, THROUGH VARIOUS LENGTHS OF BEST 3-INCH RUBBER LINED HOSE

3-inch Lines				Two 2½-inch Lines Siamesed								Nozzle Pressure Indicated by Pitot Gage
800 Feet	1,000 Feet	1,200 Feet	1,500 Feet	200 Feet	400 Feet	600 Feet	800 Feet	1,000 Feet	1,200 Feet	1,500 Feet	1,800 Feet	
87	104	120	144	33	45	56	68	79	91	108	126	20
108	128	148	178	41	56	70	84	99	113	135	156	25
129	153	177	212	49	66	83	oo	17	134	160	185	30
149	177	204	245	57	77	96	16	35	155	184	214	35
170	201	232	279	65	88	110	32	55	177	211	244	40
189	224	258	....	73	97	122	146	171	196	233	269	45
209	247	286	....	81	108	136	163	190	218	259	300	50
228	270	....	....	88	118	148	178	208	237	282	327	55
248	293	....	....	96	128	161	193	225	257	305	....	60
267	....	....	....	104	139	174	208	243	278	....	....	65
287	....	....	....	112	149	186	223	261	298	....	....	70
307	....	....	....	120	160	199	239	279	319	....	....	75
...	....	....	....	127	170	212	254	296	....	....	....	80
...	....	....	....	135	179	224	268	313	....	....	....	85
...	....	....	....	143	190	237	284	....	....	....	....	90
...	....	....	....	152	201	251	301	....	....	....	....	95
...	....	....	....	160	212	264	316	....	....	....	....	100

1⅝-INCH SMOOTH NOZZLE.—

Nozzle Pressure Indicated by Pitot Gage	Discharge, Gallons per Minute	PRESSURES REQUIRED AT HYDRANT OR FIRE NOZZLE PRESSURES GIVEN IN FIRST QUALITY 2½- AND									
		Single 2½-inch Lines						Single 3-inch			
		100 Feet	200 Feet	300 Feet	400 Feet	500 Feet	600 Feet	200 Feet	400 Feet	600 Feet	800 Feet
20	350	52	80	108	136	165	193	46	68	90	112
25	392	65	100	135	170	205	240	57	84	111	138
30	429	77	118	160	201	242	284	68	00	132	164
35	463	89	136	184	231	279	326	78	15	152	189
40	496	01	155	208	262	316	....	89	31	173	215
45	525	13	173	233	293	....	....	100	46	193	239
50	554	25	192	258	324	....	....	111	62	214	265
55	581	37	210	282	....	....	....	121	78	234	290
60	607	49	228	306	....	....	....	132	93	254	....
65	631	62	246	330	....	....	....	143	09	275	....
70	655	73	263	....	....	....	....	153	23	294	....
75	678	84	281	....	....	....	....	163	37	312	....
80	700	97	299	....	....	....	....	174	53	....	....
85	722	209	317	....	....	....	....	184	69	....	....
90	743	20	....	....	....	....	....	195	84	....	....
95	763	32	....	....	....	....	....	205	99	....	....
100	783	44	....	....	....	....	....	216	314	....	....

2½ AND 3-INCH HOSE.

Lines		Two 2½-inch Lines Siamesed								Nozzle Pressure Indicated by Pitot Gage
1,000 Feet	1,200 Feet	200 Feet	400 Feet	600 Feet	800 Feet	1,000 Feet	1,200 Feet	1,500 Feet	1,800 Feet	
134	156	37	53	68	84	100	115	139	162	20
165	192	47	66	85	104	123	143	171	200	25
196	228	56	79	02	125	148	171	205	240	30
226	263	65	91	17	144	170	197	236	276	35
257	299	74	04	34	164	194	224	269	314	40
286	.....	82	16	49	182	215	248	298	.....	45
.....	.....	91	28	65	202	239	275	331	.....	50
.....	.....	100	140	181	221	261	301	.....	.....	55
.....	.....	109	153	196	240	283	327	.....	.....	60
.....	.....	118	164	211	258	305	.....	.....	.....	65
.....	.....	126	176	226	276	326	.....	.....	.....	70
.....	.....	135	189	242	295	.	.....	.....	.....	75
.....	.....	144	201	258	314	.	.....	.....	.....	80
.....	.....	153	213	273	.....	.	.....	.....	.....	85
.....	.....	162	225	289	.....	.	.....	.....	.....	90
.....	.....	170	237	303	.....	.	.....	.....	.....	95
.....	.....	179	249	319	.....	.	.....	.....	.....	100

ENGINE, WHILE STREAM IS FLOWING, TO MAINTAIN COLUMN, THROUGH VARIOUS LENGTHS OF BEST 3-INCH RUBBER LINED HOSE

1¾-INCH SMOOTH NOZZLE.—

Nozzle Pressure Indicated by Pitot Gage	Discharge, Gallons per Minute	Pressures Required at Hydrant or Fire Nozzle Pressures given in First Quality 2½- and									
		Single 2½-inch Lines				Single 3-inch					
		100 Feet	200 Feet	300 Feet	400 Feet	100 Feet	200 Feet	300 Feet	400 Feet	500 Feet	600 Feet
20	407	63	100	138	175	40	55	71	86	101	116
25	455	77	123	169	215	49	67	84	102	120	138
30	498	91	145	199	253	58	79	00	121	142	163
35	538	06	169	231	294	68	92	17	141	166	190
40	575	20	191	262	333	77	04	32	159	187	215
45	609	35	215	294	....	87	18	49	180	211	241
50	643	50	237	325	....	96	30	64	199	233	267
55	674	64	259	....	....	105	142	179	216	254	291
60	704	77	280	....	....	114	154	194	234	274	314
65	732	91	302	....	....	123	166	209	252	296	....
70	761	06	325	....	....	133	180	227	273	....	....
75	787	20	....	....	....	143	192	242	291	....	....
80	813	34	....	....	....	152	204	257	309	....	....
85	838	47	....	....	....	160	215	270	....	....	....
90	862	61	....	....	....	169	228	286	....	....	....
95	885	74	....	....	....	178	240	301	....	....	....
100	909	..	....	....	....	188	253	317	....	....	....

2½- AND 3-INCH HOSE.

| Engine, while stream is flowing, to maintain Column, through various Lengths of Best 3-inch Rubber Lined Hose | | | | | | | | | | | Nozzle Pressure Indicated by Pitot Gage |
| Lines | | Two 2½-inch Lines Siamesed | | | | | | | | | |
800 Feet	1,000 Feet	100 Feet	200 Feet	300 Feet	400 Feet	500 Feet	600 Feet	800 Feet	1,000 Feet	1,200 Feet	
147	177	33	43	53	64	74	84	105	125	146	20
173	209	40	53	65	78	91	103	128	154	179	25
205	247	49	64	79	94	10	125	155	185	215	30
239	288	56	74	91	09	26	143	178	213	248	35
270	325	64	84	103	123	143	162	201	241	280	40
303	....	73	95	117	139	161	183	227	271	315	45
....	....	80	104	128	152	177	201	249	297	....	50
....	....	88	114	140	167	193	219	272	324	....	55
....	....	96	125	153	182	210	239	296	....	....	60
....	....	104	134	165	195	226	257	318	....	....	65
....	....	111	144	177	210	243	275	....	....	....	70
....	....	118	153	188	223	258	293	....	....	....	75
....	....	127	164	201	239	276	313	....	....	....	80
....	....	135	174	214	253	293	....	....	....	....	85
....	....	142	183	225	266	308	....	....	....	....	90
....	....	150	194	237	281	....	....	....	....	....	95
....	....	158	204	250	296	....	....	....	....	....	100

2-INCH SMOOTH NOZZLE.—

Nozzle Pressure Indicated by Pitot Gage	Discharge, Gallons per Minute	PRESSURES REQUIRED AT HYDRANT OR FIRE NOZZLE PRESSURES GIVEN IN FIRST QUALITY 2½- AND							
		Single 2½-inch Lines			Single 3-inch Lines				
		100 Feet	200 Feet	300 Feet	100 Feet	200 Feet	300 Feet	400 Feet	500 Feet
20	532	90	152	214	52	76	100	124	148
25	594	111	187	263	65	94	123	152	182
30	651	132	222	312	77	112	147	181	216
35	703	152	255	.....	89	129	169	209	249
40	752	173	290	.....	102	147	193	238	283
45	797	193	323	.....	113	163	213	263	314
50	841	214	.....	.....	126	182	237	293	.....
55	881	.....	.....	.....	138	199	260	321	.....
60	920	.....	.....	.....	150	216	282	.....	.....
65	958	.....	.....	.....	162	233	304	.....	.....
70	994	.....	.....	.....	175	251	327	.....	.....
75	1,029	.....	.....	.....	187	268	.....	.....	.....
80	1,063	.....	.....	.....	199	285	.....	.....	.....
85	1,095	.....	.....	.....	211	302	.....	.....	.....
90	1,128	.....	.....	.....	223	319	.....	.....	.....
95	1,158	.....	.....	.....	235	335	.....	.....	.....
100	1,189	.....	.....	.....	247	.....	.....	.....	.....

$2\frac{1}{2}$- AND 3-INCH HOSE.

ENGINE, WHILE STREAM IS FLOWING, TO MAINTAIN COLUMN, THROUGH VARIOUS LENGTHS OF BEST 3-INCH RUBBER LINED HOSE		Two 2½-inch Lines Siamesed								Nozzle Pressure Indicated by Pitot Gage.
600 Feet	800 Feet	100 Feet	200 Feet	300 Feet	400 Feet	500 Feet	600 Feet	800 Feet	1,000 Feet	
172	220	41	58	75	92	110	127	161	195	20
211	270	51	72	93	114	135	156	198	240	25
251	321	61	86	10	135	160	185	234	284	30
289	....	71	00	28	157	186	214	271	329	35
....	....	81	13	46	178	211	243	308·	.....	40
....	....	90	26	62	198	234	270	.....	.....	45
....	....	100	140	180	220	260	300	.....	.....	50
....	....	110	153	197	240	284	.....	.....	.....	55
....	....	119	166	213	260	308	.....	.....	.....	60
....	....	129	180	230	281	.....	.....	.....	.....	65
....	....	139	193	248	302	.....	.....	.....	.....	70
....	....	148	206	264	322	.....	.....	.....	.....	75
....	....	158	219	280	....	.....	.....	.....	.....	80
....	....	167	232	297	....	.....	.....	.....	.....	85
....	....	177	245	314	....	.....	.....	.....	.....	90
....	....	186	258	....	....	.....	.....	.....	.....	95
....	....	196	272	....	....	.....	.....	.....	.....	100

1¼-INCH SMOOTH NOZZLE.—3½-INCH HOSE.

Nozzle Pressure Indicated by Pitot Gage	Discharge, Gallons per Minute	Pressures Required at Hydrant or Fire Engine, while stream is flowing, to maintain Nozzle Pressures given in First Column, through various Lengths of Best Quality 3½-inch Rubber Lined Hose.								Nozzle Pressure Idicated by Pitot Gage
		600 Feet	700 Feet	800 Feet	900 Feet	1,000 Feet	1,200 Feet	1,500 Feet	1,800 Feet	
20	206	32	34	36	37	39	43	49	55	20
25	230	39	42	44	46	48	53	60	67	25
30	253	47	49	52	55	58	63	71	79	30
35	273	54	57	60	64	67	73	82	91	25
40	292	62	65	69	72	76	83	93	o4	40
45	309	69	73	77	81	85	93	o4	16	45
50	326	77	81	85	90	94	102	115	128	50
55	342	84	89	94	99	103	112	126	141	55
60	357	92	97	102	107	112	122	137	153	60
65	372	99	105	110	116	121	132	149	165	65
70	386	107	113	118	124	130	142	160	177	70
75	399	114	120	127	133	139	152	171	190	75
80	413	122	128	135	142	148	162	182	202	80
85	425	128	135	142	149	156	170	191	212	85
90	438	136	143	151	158	165	180	202	225	90
95	449	143	151	159	167	175	190	214	237	95
100	461	151	159	167	175	184	200	225	249	100

1⅜-INCH SMOOTH NOZZLE.—3½-INCH HOSE.

Nozzle Pressure Indicated by Pitot Gage	Discharge, Gallons per Minute	Pressures Required at Hydrant or Fire Engine, while stream is flowing, to maintain Nozzle Pressures given in First Column, through various Lengths of Best Quality 3½-inch Rubber Lined Hose									Nozzle Pressure Indicated by Pitot Gage
		400 Feet	500 Feet	600 Feet	700 Feet	800 Feet	1,000 Feet	1,200 Feet	1,500 Feet	1,800 Feet	
20	250	31	34	36	39	41	47	52	60	67	20
25	280	39	42	45	49	52	59	65	75	85	25
30	307	46	50	54	58	62	70	78	89	01	30
35	331	54	58	63	67	72	81	90	03	17	35
40	354	61	66	71	76	81	91	101	116	131	40
45	376	69	74	80	85	91	102	113	130	147	45
50	396	76	82	88	95	101	113	126	144	163	50
55	415	84	90	97	104	111	124	138	158	179	55
60	434	91	98	106	113	121	135	150	172	195	60
65	451	98	106	114	122	130	146	161	185	209	65
70	469	106	114	123	131	140	157	174	199	225	70
75	485	113	122	131	140	149	167	185	212	239	75
80	500	120	130	140	149	159	178	197	226	255	80
85	516	127	138	148	158	168	188	208	239	269	85
90	531	135	146	156	167	178	199	221	253	285	90
95	546	142	153	165	176	187	209	232	266	299	95
100	560	150	161	173	185	197	220	244	279	315	100

1½-INCH SMOOTH NOZZLE.—3½-INCH HOSE.

Nozzle Pressure Indicated by Pitot Gage	Discharge, Gallons per Minute	PRESSURES REQUIRED AT HYDRANT OR FIRE ENGINE, WHILE STREAM IS FLOWING, TO MAINTAIN NOZZLE PRESSURES GIVEN IN FIRST COLUMN, THROUGH VARIOUS LENGTHS OF BEST QUALITY 3½-INCH RUBBER LINED HOSE								Nozzle Pressure Indicated by Pitot Gage
		200 Feet	400 Feet	600 Feet	800 Feet	1,000 Feet	1,200 Feet	1,500 Feet	1,800 Feet	
20	298	28	36	43	50	58	65	76	87	20
25	333	35	44	53	62	71	80	93	107	25
30	365	42	53	63	74	85	96	112	128	30
35	394	49	61	73	86	98	111	129	148	35
40	422	55	69	83	97	111	125	146	167	40
45	447	62	78	93	109	125	140	164	187	45
50	472	69	86	103	121	138	155	181	207	50
55	494	76	94	113	132	151	170	198	226	55
60	517	82	102	123	143	163	183	214	244	60
65	537	89	111	133	154	176	198	231	263	65
70	558	96	119	143	166	189	213	248	283	70
75	578	103	128	153	178	203	228	265	303	75
80	596	109	136	162	188	215	241	281	....	80
85	614	116	144	172	200	228	256	298	....	85
90	633	123	152	182	211	241	271	....	....	90
95	650	129	160	191	222	253	284	....	....	95
100	667	136	168	201	233	265	298	....	....	100

1¾-INCH SMOOTH NOZZLE.—3½-INCH HOSE.

Nozzle Pressure Indicated by Pitot Gage	Discharge, Gallons per Minute	PRESSURES REQUIRED AT HYDRANT OR FIRE ENGINE, WHILE STREAM IS FLOWING, TO MAINTAIN NOZZLE PRESSURES GIVEN IN FIRST COLUMN, THROUGH VARIOUS LENGTHS OF BEST QUALITY 3½-INCH RUBBER LINED HOSE.									Nozzle Pressure Indicated by Pitot Gage
		100 Feet	200 Feet	300 Feet	400 Feet	500 Feet	600 Feet	800 Feet	1,000 Feet	1,200 Feet	
20	407	28	35	41	48	54	61	74	87	101	20
25	455	35	43	51	59	67	75	91	107	123	25
30	498	41	51	60	70	79	89	108	127	146	30
35	538	48	59	70	81	92	103	124	146	168	35
40	575	55	67	80	92	105	117	142	167	191	40
45	609	62	75	89	103	117	131	158	186	213	45
50	643	68	84	99	115	130	145	176	206	237	50
55	674	75	92	109	125	142	159	192	225	259	55
60	704	82	100	118	136	154	172	208	244	280	60
65	732	89	108	127	147	166	186	224	263	302	65
70	761	95	116	137	158	178	199	241	282	....	70
75	787	102	124	146	168	190	212	257	301	....	75
80	813	109	132	156	179	203	226	273	320	....	80
85	838	115	140	165	190	214	239	289	....	....	85
90	862	122	148	174	200	227	253	305	....	....	90
95	885	128	156	183	211	238	266	....	....	....	95
100	909	135	164	193	222	251	280	....	....	....	100

1⅝-INCH SMOOTH NOZZLE.—3½-INCH HOSE.

Nozzle Pressure Indicated by Pitot Gage	Discharge, Gallons per Minute	Pressures Required at Hydrant or Fire Engine, while stream is flowing, to maintain Nozzle Pressures given in First Column, through various Lengths of Best Quality 3½-inch Rubber Lined Hose.								Nozzle Pressure Indicated by Pitot Gage
		200 Feet	400 Feet	600 Feet	800 Feet	1,000 Feet	1,200 Feet	1,500 Feet	1,800 Feet	
20	350	31	41	50	60	70	80	94	109	20
25	392	38	51	63	75	87	99	118	136	25
30	429	46	60	75	89	03	18	139	161	30
35	463	53	70	86	103	120	136	161	186	35
40	496	61	79	98	117	136	155	183	211	40
45	525	68	89	110	131	152	173	205	236	45
50	554	76	99	122	145	168	192	226	261	50
55	581	83	108	133	158	184	209	247	284	55
60	607	90	117	144	172	199	226	267	308	60
65	631	97	127	156	186	215	244	289	....	65
70	655	105	136	167	199	230	262	309	....	70
75	678	112	145	179	212	245	279	....	....	75
80	700	119	155	191	226	262	297	....	....	80
85	722	127	165	202	240	278	316	....	....	85
90	743	134	174	214	254	294	....	....	....	90
95	763	141	183	225	267	309	....	....	....	95
100	783	149	193	237	281	....	....	....	....	100

2-INCH SMOOTH NOZZLE.—3½-INCH HOSE.

Nozzle Pressure Indicated by Pit t Gage	Discharge, Gallons per Minute	PRESSURES REQUIRED AT HYDRANT OR FIRE ENGINE, WHILE STREAM IS FLOWING, TO MAINTAIN NOZZLE PRESSURES GIVEN IN FIRST COLUMN, THROUGH VARIOUS LENGTHS OF BEST QUALITY 3½-INCH RUBBER LINED HOSE.								Nozzle Pressure Indicated by Pitot Gage
		100 Feet	200 Feet	300 Feet	400 Feet	500 Feet	600 Feet	800 Feet	1,000 Feet	
20	532	33	44	55	65	76	87	109	130	20
25	594	41	54	67	80	93	106	133	159	25
30	651	49	64	80	96	111	127	158	189	30
35	703	57	75	93	111	129	147	183	219	35
40	752	65	85	105	126	146	166	207	247	40
45	797	72	95	118	140	163	185	231	276	45
50	841	80	105	130	155	180	205	255	305	50
55	881	88	116	143	170	197	225	279	....	55
60	920	96	126	155	185	214	244	303	....	60
65	958	104	136	168	200	232	263	....	....	65
70	994	112	146	180	214	248	282	....	....	70
75	1,029	119	156	192	229	265	301	....	....	75
80	1,063	127	166	205	243	282	....	....	....	80
85	1,095	135	176	217	258	299	....	....	....	85
90	1,128	143	186	229	272	....	....	....	....	90
95	1,158	151	196	241	286	....	....	....	....	95
100	1,189	158	206	253	301	....	....	....	....	100

As fire-fighting is not and probably will never be an exact science, the following rules are given as a guide to men who may not have had the advantage of a higher mathematical education. It is not claimed that these rules are absolutely correct but they are sufficiently accurate to be of practical value and can easily be understood by any man in the fire service of any country.

1. For placing of ladders.

In placing an extension ladder, the hub of front wheel should be placed opposite centre of window, and when placed at proper distance from base of building, extended so as to be 1 ft. and 3 inches over the window sill and 1 ft. from the building. The foot of ladder should be placed according to following rule:

Divide length of ladder in feet by 5 and add 2.

The point of ladder should not rest against the building, but should be about one foot from it. ·

2. For obtaining discharge in gallons, when nozzle pressure is given.

Square diameter of nozzle.

Multiply above product by the average barometric pressure 29.9.

Then multiply this by the square root of the nozzle pressure.

Example:

What would be the discharge of a 1½-inch nozzle working at 100 lbs. pressure?

1.5×1.5 equals 2.25.

2.25×29.9 equals 67.275.

The square root of 100 is 10.

67×10 equals 670 gallons discharge.

3. For determining number of lengths of hose required when location of fire is known in building.

1 length of hose for each story and 1 length for good measure.

Example:

How many lengths would be required to stretch to the 7th story.

7 lengths for height and 1 length to cover floor or 8 lengths.

4. For obtaining pressure required to deliver water at any given point in a stand-pipe.

Multiply height in feet by. constant .434 and add 25 lbs. for friction. (5 lbs. for entry in siamese connection, 10 lbs. for loss in passing through swing check, and 10 lbs. for outlet valve.)

Example:

What pressure would be required to deliver water to outlet on 12th floor or at 150 feet?

150 × .434 equals 65 lbs.

65 plus 25 lbs. for friction equals 90 lbs. pressure required to deliver water at that point. (To this must be added the nozzle pressure desired.)

5. For obtaining pressure exerted by water dropping from a tank through a stand-pipe at any given point.

Divide drop in feet by constant 2.31.

Example:

What pressure would there be at an outlet on ground floor from a tank located 150 ft. above? (150 divided by 2.31=65 lbs.)

6. For obtaining the friction in hose.

Friction loss in hose is controlled by the amount of flow, and is calculated as of 100 ft. lengths.

Loss in 3½" hose.

For a flow of 500 to 1,200 gallons per minute.

The loss for first 500 gallons is 9.5 lbs. and for each 10 gallons over up to 1,200 add .6 of a lb.

Example:

What would be the loss in 100 ft. of 3½" hose discharging 800 gallons per minute?

For first 500 gallons would be 9.5 lbs. and for the 300 gallons over (which is 30×10), should be added 30×.6 or 18 lbs. to the 9.5, which would be 27.5 lbs. loss in friction.

Loss in 3″ hose.

For a flow of 200 to 400 gallons per minute.

The loss for first 200 gallons is 4 lbs. and for each 10 gallons over, up to 400 add .5 of a lb.

Example:

What would be the loss in 400 ft. of 3″ hose discharging 310 gallons per minute?

For first 200 gallons the loss would be 4 lbs. and for the 110 over (which is 11×10), should be added 11×.5, or 5.5 lbs. to the 4 lbs., which would be 9.5 lbs. for each 100 ft., and for 400 ft. would be 4×9.5, or 38 lbs. loss.

For a flow of 400 to 700 gallons per minute.

The loss for first 400 gallons is 14 lbs. and for each 10 gallons over, up to 700, add .8 of a lb. (figured as above).

Loss in 2½″ hose.

For a flow of 200 to 400 gallons per minute.

The loss for first 200 gallons is 10 lbs. and for each 10 gallons over, up to 400, add 1.3 lbs.

Example:

What would be the loss in 300 ft. of 2½″ hose discharging 350 gallons per minute.

The loss for first 200 gallons is 10 lbs. and for the 150 gallons (which is 15×10), should be added 15×1.3 or 20 lbs., to the 10 which would be 30 lbs. for each 100 ft., and for 300 ft. the loss would be 3×30 or 90 lbs.

7. For effective reaching distance of fire streams either vertical or horizontal.

To do efficient service allow 1 lb. in pressure for each foot in distance.

8. For finding horsepower of a fire engine.

Multiply the area of the piston by the steam pressure in pounds per square inch; multiply this product by the travel of the piston in feet per minute, divide this result by 33,000, and .7 of this quotient will be the horsepower.

Example:

What would be the horsepower of a fire engine with cylinders of 8″ diameter, 80 lbs. steam pressure, with a stroke of 6″ and traveling 200 revolutions per minute?

8×8 equals 64.

64 × .7854 equals 50.2656 area of piston.

50.2656×80 equals 4021.2480.

6×2 equals 12″ or 1 ft. travel of piston for 1 revolution.

1×200 equals 200 ft. travel of piston per minute.

4,021×200 equals 804,200.

804,200 divided by 33,000 equals 24.37.

24.37×.7 equals 17.059 horsepower.

9. For obtaining pump capacity.

Square the diameter of pump cylinder.

Multiply above product by travel of piston in inches per revolution.

Then multiply this result by constant .0034.

This will give the displacement of the pump in gallons per revolution, less the displacement of the plunger rod, which must be deducted to get the net displacement.

To get the displacement of plunger rod.

Square the diameter, then multiply by the travel of the piston for ½ revolution, then multiply this by the constant .0034.

Example:

What is the capacity of a pump having a 6″ cylinder, stroke 9″, plunger rod 1½″, traveling 300 revolutions per minute?

6×6 equals 36.

9×2 equals 18 inches travel for 1 revolution.

36×18 equals 648 cubic inches.

648×.0034 equals 2.2032 gallons capacity per revolution, less the displacement of the plunger rod.

1.5×1.5 equals 2.25.

2.25×9 equals 20.25.

20.25 × .0034 equals .06885, displacement in gallons of plunger rod.

2.2032 less .06885 equals 2.13435 gallons, net displacement per revolution.

2.13435×300 equals 640 gallons, capacity of pump.

CPSIA information can be obtained
at www.ICGtesting.com
Printed in the USA
LVOW04s1035160116

470920LV00021B/152/P